People Say The Coolest Things...

"I started writing for variety
and went on to receive an E
writing a documentary. I ow
Access."

                    - Marilyn Brennan
                    Emmy nominated writer

"Now that I'm out of the ring, Public Access is a
great way for me to reach my audience. And my
friend Glenn Darby knows Public Access!"
                    -Ray "Boom Boom" Mancini
                    former lightweight World Champion Boxer

"I had a great time being a guest on Pop Bop! --
Glenn had a totally creative way of approaching his
show. It was less hype, just seemed like a fun time
hanging out with a friend talking about music with a
camera rolling . . . and he even baked me a cake!"
                    -Starr Parodi
                    Composer, keyboardist

"Glenn has been fabulous in helping us with our
shows at the studio. He knows how to get through
the maze of details, requirements and problems."
                    -Carole Dean
                    President, Studio Film & Tape

"Glenn has the skill, creativity and knowledge
necessary to teach people how to have their own
successful TV show. We at Filmmakers have
always benefited from his experience, and you will
too!"

                    -Michele Mills
                    Producer, Filmmakers

HOW TO STAR IN YOUR OWN TV SHOW FOR $50 OR LESS!

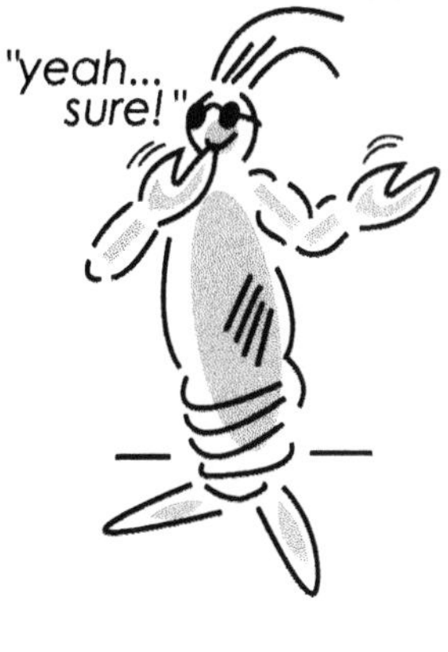

Big Red Barn Productions
1247 Lincoln Blvd #158
Santa Monica, CA 90401 USA

ISBN 0-9669328-0-3

Library of Congress Catalog Card Number:
99-94514

www.howtostar.com

Illustrated by

ALLAN
CAMERON

graduate of
U. of Wisconsin
& Chouinard Art Inst.

whose work has appeared in
publications for Getty Inst.
UCLA, UAW, L.A. Unified
School Dist.,Kaiser Fdn.,
Am.Inst.Family Relations,
Ward Ritchie Press & others.

A portion of the profits from the sale of this book
will be donated to local charities to encourage
and further self-expression.

This book is dedicated to all the interns, guests and producers I've been fortunate enough to work with and share knowledge and creativity with.

And to the belief in Access TV for self-expression.

So rock the boat, rock the vote and wake up the world!

Oh, and just a few last minute thoughts...

I just got the laser copy pages back from the illustrator and wanted to let you know about our website:

www.howtostar.com

We will be constantly updating and adding more information. Please send us your tips so we may pass them on to everyone, and we'll try to get it in the next book printing. Our current email is glennheidi@earthlink.net., or glennheidi@howtostar.com.

And a last minute note: I've done everything possible to make sure all the information is current and correct. If you find a mistake or change in information, please let us know.

Stop waiting and start creating!

INTRODUCTION

Hello, Curious Person. This is the part where I get to tell you about myself and how this book came about. And maybe even impress you enough to not only buy it, but use it. Okay, here goes.

I was born in Paris, France. My dad was an army sergeant -- guess that makes me an "Army Brat." One of the best memories I have is at age nine, when I would surf every day after school at Waikiki Beach in Hawaii. Very cool!

Fast Forward

While still a teenager, I began making cakes for rock 'n roll bands/musicians, who came through Philly (and New York). Who? Okay, I'll drop some names: Bruce Springsteen, Fleetwood Mac, Elton John, Billy Idol, The Rolling Stones, Billy Joel, Elvis Costello, Tom Petty, David Bowie, Poison, Dokken, Judas Priest, Iron Maiden, Metallica, Motley Crue, Foreigner, The Jacksons, Genesis, Def Leppard, Queen, The Knack -- enough already? Tina Turner, Anita Baker . . . and, I got to meet many of them, too!

Fast Forward Again

While working in the mailroom (I know it sounds like a cliche -- but even David Geffen, an inspiring self-made billionaire and philanthropist, did too) at *Billboard Magazine*, I started my own TV show, where I got to meet bands and interview them. Names? Phish, Everclear, Candlebox, Dramarama, Mavis Staples, Kiss, Jonatha Brooke, Ween, Venice. Okay, so some aren't as well known as the others, but they were all great to interview.

Forward -- Not So Fast

I'm a director! No, I didn't go to film school -- actually, I have an Associate's degree in Natural Science and a Bachelor's degree in English. In the five-and-a-half years I spent in the studio, I've directed about two thousand shows. What kind of shows, you may ask? That's what this book is all about.

During my last year, I started writing a TV sitcom (treatment, bible, etc., which is currently in development) about what it's like to be at an actual cable TV studio, based on the people I was working with. And when that was finished, I realized how much information I had to share with people who wanted to do their own TV show.

I figured it this way -- I've got all this experience, where I've taught people how to do everything in a studio: run cameras, shoot on location, edit, technical direct, light, direct three camera shows, produce shows, etc. -- so why not write a book. And now you have the results.

By the way, did you check out the "blurbs" I got from other people?

The reason I'm telling you all this information is that most of my life, I never thought I could do something until I tried it. The cakes, the music show . . . and that's exactly what I want you to do!

Oh, one thing before you begin the first chapter -- I should mention some people who really, really helped me out: first my wife, Heidi, who truly loves me and supports my creativity; my daughter, Zaidee, who gives me reason to live and learn; my "Mom" -in-law, Marilyn, who painstakingly helped me edit this book; and Tony Robbins, who is the most passionate of all coaches -- Yumbo!

So hang on tight, my friend, we're about to go on an adventure together.

-- *Glenn*
Santa Monica, California

"Wow!"

TABLE OF CONTENTS

CHAPTER   PAGE

GETTING STARTED

First, let me say thank you for picking up this book, reading a little and even buying it. You are taking an action towards fulfilling your dream of having your own TV show. I, too, am fulfilling my dream by guiding you through the process. So thank you.

I promise to give you the right information and a way to organize that information into a step-by-step process, that will have you starring in your own TV show. Plus, I'll tell you how that show could be seen by potentially millions of viewers -- for little or no money. Sound like one of those late night infomercials? Trust me, it's not!

Let's Get Started!

The first thing you need to determine is the kind of show you want to have. And the best way to figure that out, is to do something you like to do. Ask yourself now, what is it you like to do?

Remember, you are going to be putting a lot of work into this show, so pick a subject that is exciting to you. Don't worry, you can always change it if your first idea doesn't work out. For example, one guy had a show about sports, where he and some friends would sit around talking about a variety of sporting events, and to be honest, it was pretty typical and dull. Then one day I saw him hosting his own psychic type show, and he never went back to sports! There's no law saying you can only do one type of show.

Now Lets See What You Like To Do

Talk: Do you like to talk to people and find out what makes them who they are, like Barbara Walters? Do you want to review movies with a friend, like Siskel and Ebert? Would you like to talk to authors of books, or comic books?

Demonstration: Can you demonstrate a skill and teach it to the audience? Are you good in karate or ballet? Tennis or hockey? Can you fix cars or build models? Play drums or piano? Are you good with plants and flowers? Do you get pleasure out of teaching people?

Performance: Do you like to perform? How about making your own music video? Are you in a band or like to sing? Maybe you're an actor and need a tape to showcase your talent. Or a comedian who wants to perform a stand-up routine.

Variety: Perhaps you could host a show with a little variety. You could interview people, have bands perform, have guests demonstrate skills, etc.

Just You: Can you do a show, by yourself, talking to people at home? What about wanting to make people feel good about themselves? Would you like to inspire them by reading scriptures or guiding them through meditation? Can you read Tarot cards or are you psychically in tune with the Universe? Horoscopes are fun. Do you know a lot about them? Are you a financial planner and have some tips for people? Are you tired of those music channels that never seem to play enough music videos? You can host your own music video show (I even have a list of people to contact in the reference section -- but keep reading).

No Matter What You Like To Do, There Is An Audience For It

Find out what you're comfortable with, and what you are also good at. If you're good at something, it'll be easier to do. If you're not good at talking to people, then you won't have fun and your guest won't either -- not to say you can't develop your skills, but if you like to get crazy and dance around by yourself, then do that first! The main thing is to have fun, because when you're having fun, the audience can't help but have fun.

While you're thinking about what to do, you should also keep in mind who your audience is. Are you doing a show for young kids, teens, the twenty-somethings, baby boomers, retired people? Not only knowing what your subject is, but who are you going to relate to, will be helpful in your show's appeal.

For example, if you want to reach teens, then check out what MTV is playing. Or watch the Saturday TV line up. You'll notice even the commercials are geared towards the teen audience and have a particular style, usually fast cuts and loud music.

So How Are You Going To Keep Them Watching?

Nowadays with cable companies offering a hundred channels or more, you don't have much time to grab their attention. I'll tell you how to edit (Chapter 17), and use special effects, so you can put together a show that will get their attention. Plus, we'll talk about ways to find interesting (and sometimes famous) guests. But mainly, we will be focusing on you and your own unique personality.

Kids

What about a show designed for kids? My daughter watches *Sesame Street, Teletubbies, Catdog* and *Rugrats,* which have a simple formula -- fun characters, that kids can relate to. She sings along with the songs and dances around on the floor while watching. It's because they relate to her on a level she can understand (actually I like the shows too and sing the songs with her).

A woman came into the studio where I was working and did a show dressed up as a pink bear, and entertained four kids for the entire show. You couldn't help but watch her.

Some other children's shows had kids singing and dancing, playing videos, talking about the news, exercising and even acting (it is L.A.).

Step By Step

So take a moment and come up with a simple subject you'd like to have a show about.

The next thing is to figure out what format your subject fits into. There are basically five types of shows: (1)*Talk,* (2) *Demonstration,* (3) *Performance,* (4)*Variety,* and something I call, (5)*Just You.*

In the following chapters, I'll be giving you examples of TV shows I've worked on, or ones that I'm very familiar with -- you'll see some of the names of these shows at the top of various chapters. It should help you with some ideas for your show. Okay? Let's go to the next chapter.

TALK SHOW

Shut Up and Eat; Unzipped; Heart to Heart; AIDS Vision; Filmmakers; Healthstyles; Art Fein's Poker Party; Pop Bop!; Looseleaf Report; Camp Out with Mat and Jeff.

Did You Find An Interesting Subject?

Yes? Great, let's talk about how to do it. If not, then read on for inspiration and ideas. This is a step-by-step process, where we're building your show together. Okay, let's do it.

The Most Popular Format

In this chapter we'll talk about the most popular format: the Talk Show -- sometimes referred to as talking heads (not to be confused with the group). While working in a TV studio, I found that talk shows comprised about 50 to 75 percent of all shows we did, yet it varied from show to show. Some had different numbers of guests and some had roll-ins (video segments). But they all fit into a simple interview type format where the host asks the guests questions for the entire show. Typically, it would be as if we the audience were eavesdropping in on two people's conversation -- without them knowing it. Rarely would the host (or guest) look into the camera, except to open or close the show. The idea was to invite the audience in so they could watch and listen to the interview.

Don't Get Discouraged

If you mentioned your idea of having a talk show to someone, they might have responded, either with, "You can't do that," or "Oh no, not another talk show!" But don't let them discourage you, they're probably just jealous that they don't have the courage to go after their dreams. My mother still wants me to be a doctor or lawyer -- her dream was to have a son who was a well respected "professional." Well, that isn't my dream! Don't listen to anybody who tells you what you should do with your life.

Back to your show and dream. There are (and were) famous talk show hosts like Oprah, Rosie O'Donnell, Leeza, Dave, Sinbad, Magic, Jay, Howie, Donny and Marie, Roseanne, Dennis Miller, and by the time you read this there might be a few more and probably even a few less. So there's no reason why you can't have a popular talk show too. You might not be on a big network, but there are ways to be on TV in many cities nationwide. And maybe even get on a big network, or cable channel someday. TV executives are always looking for the next "hot thing" and your TV show can demonstrate what you are capable of doing.

But please, don't do this show with the idea that after your first one, the executives at NBC, CBS, FOX or ABC are going to call you up and put you on their network. It could happen, but it usually doesn't work that way. You have to build up a reputation and following. Then when you get popular, people can't help but inquire about your show.

Everyone Starts in the Chorus

You must realize that everyone has started somewhere and worked their way up. All talk show hosts started by doing something else. Conan O'Brien was a writer, Jay and Dave comedians, Oprah did local news, Leeza was a reporter, Sinbad was on Star Search, Magic Johnson played basketball, Dennis Miller was on Saturday Night Live, and Mr. Pete was a waiter.

Who is Mr. Pete?

Who is Mr. Pete? He started his talk show on a local cable channel -- just like you're going to do (it was the first show I worked on as an intern -- more on that in Chapter 18). Then the Los Angeles local UPN affiliate picked up his show and started producing it. I think it lasted about thirteen weeks, then got canceled -- but that's better than a lot of national sitcoms! Later on I saw him on the FX channel as a host.

Starting Small Can Pay Big Dividends

Okay, you know you have to start small and you might not be able to talk to celebrities right away, but there are many other interesting topics to talk about, and interesting people to interview. Let's start with something you already know. How about music? How do you get Elton John, Aerosmith, Puff Daddy or even Whitney Houston to come on and talk with you? Sorry wrong book!

Let's take it easy -- keep the dream of interviewing Elton John alive, because you can make it happen. Try something simple like interviewing a friend of yours, who also likes music, or even better is a musician. If you don't know a musician, I'm sure a friend of yours must know one -- everyone knows a musician. The reason is that you will be more comfortable talking to a friend (or friend of a friend) -- it can be very scary in a TV studio, your first time, with all those hot lights on you, and if you freeze up or panic, then you won't be embarrassed in front of a stranger, and even worse get a bad reputation. Then no one will want to talk to you -- just kidding!

And even if you do make mistakes and get nervous, there's still no reason to give up. The only way you can get better is by doing it over and over again -- which you will do. Right? Good. Seriously, you'll be hooked on it like you never knew it was possible to be addicted to something. And it's good for you!

When I was doing my music show, I eventually got to a place where record companies were calling me and asking if I'd like to go backstage and interview their bands. But it took me a while -- and a lot of mistakes -- to get a good reputation. And you can do this too!

So you call up a friend and coax them into being your first guest on your debut show. They'll be thrilled you thought of them. We'll talk more about how to get those really good guests in Chapter 11, but for now let's keep reading.

Checklist Time

You should have a subject to talk about -- in this case, music. Your format is a talk show, and you have a guest to interview -- a friend. But what do you call it? Uh-oh, there's goes that heart pounding thing again! No problem, you have me on your side!

Name That Show

Try something simple; for example, *I like Music*, or *Music for Everyone*. Just make it simple and personal. You can even name it after yourself -- *Talking Music with Mary* or *Bob's Music Show*. Mine was called *Pop Bop!* --don't ask, it just happened one day (I went through so many silly names just like a lot of bands do).

Here are some music shows I've seen where I worked, *Offbeat, Eat to the Beat, Art Fein's Poker Party, Not Just Another L.A. Music Show, Rock in a Hard Place, Dre's World, Cool Mess, Reggae Train.*

If you're not doing a music show, here are some other local talk shows I've seen: *Change Your Voice*

Change Your Life, Tinseltown Queer, Skip E. Lowe Looks at Hollywood, Colin's Sleazy Friends, Hollywood Beat, UCLA Bruin Talk, A Diet To End Diseases.

Notice how each one is unique? You can also tell what the show is about from the title, or at least get a feel for what it is about. For inspiration, let's look at some more topics and shows.

What's In a Name?

Want to talk about the latest in AIDS health care? *AIDS Vision* brought on experts to talk about advances in the treatment of AIDS and was sponsored by APLA (AIDS Project Los Angeles). The host was Allison Arngrim, the former Nellie on the 1960's show *Little House on the Prairie.* Yes, she's all grown up and hosted her own TV show. At one point the show *AIDS Vision* was broadcast around the country.

Allison also did great work for AIDS organizations in educating people about the disease. Plus she had a stand-up comedy act and could be found at local clubs and AIDS fundraising events making people laugh.

Want to ask a unique mother and daughter about their views on current issues -- especially those dealing with women? Call up *Unzipped.* You could talk to them live (more on Live shows in Chapter 19). They would find the most ridiculous articles about people that you would think were fiction, but were true. Then they'd invite people to call-in and

11

give their opinions. Topics such as the O.J.
Simpson trial sparked a lot of callers.

Need to say something about gay rights?
Tinseltown Queer did just that. He had people
come on and talk about gay and lesbian issues.

Then there was Tom Litner who had a show called
Go Tell It, which assured gay people that God loves
everyone. And along the same theme was one of
my favorites, *Sister Paula,* an older transvestite
preaching about God and gay rights. She was great!
You couldn't help but love her.

More Than A Mouthful interviewed gay and lesbian
artists -- both music and paint mediums. The host,
David, also had a dating game show (*Tricks*) that
was very popular. He then decided to have a talk
show focusing on the creative side of the gay and
lesbian community.

You Don't Have To Specialize

You don't have to confine yourself to a certain area.
I was fortunate enough to interview Chastity Bono
when her band *Ceremony* put out their first album. I
knew she was a lesbian, but my music show wasn't
about the issues, it was strictly a music show. We
had fun talking about the record and her music
career and not about issues.

The Seinfeld Connection

There was a producer, Colin, who has had
actress/comedienne Janeane Garafalo (*Ben Stiller*

Show, Truth About Cats and Dogs, Reality Bites) as a guest. He has also had an article written about him in *Rolling Stone* magazine, and even had a small speaking role on *Seinfeld,* where he *was* billed as The Sleazy Guy, just like his show.

On his talk show, *Colin's Sleazy Friends,* he interviewed people who worked in adult movies. He told me once, "I'm just an idiot who likes porn." Usually when he taped a show, the control room would fill up with the installers who worked for the cable company, and the male interns would ask for autographs and pictures with the women. It just goes to show you how one subject can be popular with a certain audience.

Sometimes, when people ask me where I've worked and I mention the studio, they ask me about Colin -- he is very popular, but again, with a certain audience.

What About Sports!

What ABOUT sports? Let me tell you about *UCLA Bruin Talk* -- a show I directed exclusively for four years. It was (and still is) a show that highlighted the school's sports teams. They had two student hosts, Michelle and Greg, who would interview up to four athletes per show. It was arranged into two segments per show. The coaches would also come on and talk about the athletes, while they were sitting next to them, blushing. Every Sunday around 4:30 p.m. you could tune and hear about the Bruins, whether it was softball, basketball (the National Champions), football, swimming, tennis (Pete Sampras's sister was a coach) or water polo -- they would talk about it.

13

What About Diet?

Did you know that your diet can help you fight diseases? Stig and his wife would tell you how you could use food to fight off cancer and other life threatening ailments on their show, *Diet to End Diseases.*

Is there some information you can provide to help people too?

Dare to Be Different

How about something completely different. *Carl Bradley Pitches TV Show Ideas,* was hosted by Carl Bradley, who was actually Brad Slaight. His "host" character was so realistic, that people thought Brad was Carl, and Carl was Brad. His show was centered around the idea that networks spend millions of dollars to develop sitcoms (they actually do), but he could do the same for $35.00.   How? What he would do is tape a sitcom idea -- about five to seven minutes long -- on location (see Chapter 16) and use three of those segments for his show. He would also have an actor or two, from the skit, talk about what they did.

And Brad, I mean Carl, would always have me (the director) in the opening. One time his friend Mark, who was a Sam Kinison look-alike, was in the control room with us for a bit about Sam-Kinison-in-Public-Access-Hell. It was hilarious! And it was fun being typecast as "The Director."

Maybe you and a friend have an opinion about everything (and who doesn't). Well, these two guys Mat and Jeff would talk about anything. They both worked at Starbucks and would tell funny stories about people who came in -- some were celebrities. They also reviewed and rated movies on a scale of shoes: brown leather pump with four-inch heels! They would just have fun, without caring what people thought about them, and that is the key to doing your own show -- not to care what others think! Thanks, guys, for keeping that feeling alive.

Music Videos - Love Them or Hate Them

Maybe you're pro-music but anti-music video. *Art Fein's Poker Party,* was seen around the country. Art told me once his show tapes "had a life of their own," being passed around from city to city. He would interview the lesser known bands of today, pioneers of rock music and your classic blues artists. I remember recently at the studio, seeing Doug Figer of The Knack sitting in the green room, waiting to go on Art's show. I just had to tell him I thought The Knack was cool. He smiled.

Art doesn't play music videos -- but when he plays music (from CD's), what you see on your TV is a blank screen with a disclaimer -- "music doesn't have to be seen to be good." He's got a good point. He lets you use your own imagination as you listen to the music. He was unique compared to all the other pro-music video shows.

Kvetch

There was a woman, Laura, who came down to the studio to do a show called *Stretch n' Kvetch*. She made a big hit with one of my fellow directors because her husband was executive producer on a Star Trek TV series -- I didn't have a clue what he was raving about. But, being that we were in Los Angeles, his adoration for the show, and helping her out, he got a one time acting part on the show. Favors are the way a lot of people get work and popular with their shows -- especially in Hollywood.

For example, I worked on her show a few times and it was fun -- in return she made a cameo in my first video movie (it was quite an experience and experiment). Her show was basically a talk show where she and her co-host would talk (kvetch) then go down on the floor to stretch, and kvetch more -- which made it slightly different, but still a talk show.

Love

How about looking for the love of your life -- or just a date? We had women acting like matchmakers on a few shows. They would talk about how women over a certain age could find the right men. These two women, who were in their forties and fifties, had more energy and zest for life than woman half their age. And I heard that they were dating like crazy because of the show! (Disclaimer: I make no guarantee that the show can be used for getting dates, but if you do -- have fun!)

16

Let's Summarize

A talk show is basically two people, sitting in chairs, by a table with some plants in the background, in a studio, talking about their favorite subject, for half an hour (actually twenty-eight-and-a-half minutes -- but we'll talk about that later).

Okay, that's the end of this chapter. Take time and get everything I talked about ready: (1) theme; (2) format, it is a talk show; (3) guest, we mentioned a friend that you're comfortable with. You can skip now to the chapter titled "Where to Go," or take time and get inspired by finding out about the other formats, like . . .

COWABUNGA!!

DEMONSTRATION

Hollywood Yogi; Salazar Makeovers; Looking Fabulous; Practical Self Defense; Creole Cooking; Everyone Can Exercise; Simply Cooking, with Doris Romeo; On Tap, with Diane Davisson; Scene Study with Doug Warhit; Weights, with Ron Kennington; Alyssa's Raw and Wild Food Show.

This chapter is about demonstrating a skill to the audience, which they can follow in a step-by-step way. As you'll see, it can be about anything you are skilled at and want to teach to others.

Music

A former intern, Brian, was a piano player and wanted to teach people to play the piano. So one day we booked some studio time, he brought in a portable piano and we did a show. It was fun as we learned some chords, fingering and progressions.

You could do the same with a guitar, or even a drum set. There was a show called *Drumst6,* that

18

gave lessons on drumming. One show I directed had a guest studio drummer named Mr. Porcaro -- his sons played in the band, Toto. It was very easy to learn from him, as he demonstrated various techniques for playing a full drum set. I even got a pair of autographed drumsticks after the show.

What Can You Teach People? Keep Thinking . . .

So you want to learn how to tap dance? Diane Davisson would show you step-by-step how to do it (just bring your tap shoes). Her show, *On Tap*, had her flanked by two dancers as she went through the basics -- heel toe, ball change, etc. You got to see close ups of her feet so you could get the moves down. By the end of the show you'd be doing a full dance routine. Her key to teaching was her skill and passion for tap dancing.

Maybe you want to learn how to ballroom dance. Then you would've tuned into John Cassese's *The Dance Doctor* show and learned to tango or fox-trot, or any dance that was popular. This was his profession and he was excellent at it (I've heard that some Hollywood producer was considering making a movie about his life).

Do you like vegetarian cooking? You don't need to tune into the Food Channel -- although my wife and I like those Hot Tamales, and sometimes watching Emeril is fun. Instead, you could watch Doris Romeo. She would bring in her ingredients, a few pans, a hotplate and cook up three delicious dishes. You would see her chop it, blend it or mix it, then put it in the fry pan and cook it up. And afterwards she'd feed it to the crew.

Along the same style was *Alyssa's Raw and Wild* food show, which had a different way of demonstrating her technique. She built a self-contained, rolling-cooking- island-table, that had a mirror on top -- so you could get the overhead view of what she was doing. All her food required no cooking. Besides being healthy it tasted good!

No matter what you're good at, with a "Demonstration" show, you could teach people something within half an hour. Think about it, while we look at more shows and try to get you inspired.

Speaking of Inspiration . . .

Maria Serrao was in a car accident as a little girl, and the result was her being confined to a wheelchair. But she didn't let that stop her from starring in her very own exercise show called *Everyone Can Exercise.* She would come into the studio with a partner and go through a full aerobic, exercise and weightlifting routine. (People would always comment on her big biceps!)

Besides being an attractive woman, it was her personality that kept people watching. Eventually she was able to develop a mass market video that she sold nationally on QVC. All because she took the first step and decided that she was going to star in her own TV show. I've had the pleasure of directing some of her shows in the studio and on location -- and must say she is very inspiring to watch, and fun to be with.

20

Some More Ideas

Are you a Personal Trainer? Grab your weights and head down to the studio. Just like Ron Kennington's show, *Fitness and Weights*, you too can set up an exercise program for people to do at home while watching TV. And you might be able to go into business for yourself.

How about computers? Do you know how to work a computer very well? Why not do a show where you demonstrate to people the basics of computers? They have all those books sitting on shelves for people to read and learn from, but how about showing them instead? There are so many software programs out there and people just don't understand how to work them -- I know because I still don't understand as much as I'd like to. A lot of us could use your help!

So have you been inspired yet? No? Okay, I'll give you some more examples.

And Some More . . .

Want to learn how to sing? The *Singing with Roger Burnley* show could teach you how. He would bring on students and demonstrate how each person's style was different and how each person had special needs to develop their voices. He went through some exercises and taught the dynamics and physical attributes of voice control. As I mentioned before, it had to do with personality in the way he demonstrated the material -- you felt comfortable.

The show started with an idea by an intern who was being coached by Roger. Next thing you know, he was in the studio in front of a microphone teaching people how to sing. We watched beginners and advanced students as they took their turn listening to his instructions and improving their singing.

How about your speaking voice? Not happy with it? You could tune into Dr. Mort Cooper's show, *Change Your Voice Change Your Life*, and he would demonstrate how to speak with more confidence. He also promoted his various books on the subject of voice conditioning too.

Are you an acting coach? Doug Warhit, who has been in movies and on TV shows, (I just saw him the other night on *NYPD Blue*) had his own scene study show. He would have his students read a two-person short scene, then critique it and have them do it again, which you could see was better. As a coach, he would demonstrate to viewers how to make a scene better -- whether it was cold readings (you have to know this if you want to be an actor) or rehearsed. His students liked it and gained from the experience. Best of all, it gave them an opportunity to be on TV.

But Exaotly How Do I Go About Doing A Demonstration Show?

Okay, now that I've got you inspired, let me help you set it up. Let's say you are a hair stylist and want to demonstrate your technique. Here's exactly how to go about doing it.

I've worked on three different hair styling shows, and have found it best to have a model, fully made up and ready to go. All you need is your scissors, some hair spray and a blow dryer. You won't be able to wash their hair -- but you can bring a spray bottle to wet their hair. Break it down into segments: first you start cutting, as the cameras follow your hands, then you style it, dry it, and that's your half-hour show. See how easy?

You can even do a before and after shot. Have the director take a left-side, half screen, freeze-frame shot of your model, and store it for the end of the show (on a TBC). Then at the end of the show, ask the director to use a split screen and use the model's finished look to fill up the other half -- right-side -- of the screen.

While James Romano's, *Looking Fabulous* and Mogi's *Pro Steps* would do a straight forward haircutting show, Salazar (clients included Yoko Ono and E.T.'s Mary Hart) on his *Show Makeovers* would use video tape roll-ins to demonstrate how he cut, colored and styled hair on previous models. He would sometimes mix it up by using a live model, with roll-ins to show a wide range of styles. One lady actually had her hair made into a hat!

Now That You're Inspired . . .

You have to be inspired by now! I know you can do something fun and share it with the audience. Right? Great!

So get out those dance shoes, grab your tennis racket, dust off that guitar, pack your pots and pans, pick up the yoga mat and practice your skills, because you're the star in your own TV show.

Reminder

You only have twenty-eight minutes of show time, and you only have about two hours to set up, tape and wrap. Plan it out carefully and make it so people can learn in the allotted time, but also keep them wanting more -- say, ". . . and next time, I'll show you . . ."

Checklist

- Subject -- Your Choice! Something you want to teach.
- Format --Demonstration
- A Plan -- See Chapter 12.

And -- plenty of enthusiasm.

PERFORMANCE

Flamenco Guitar with Alberto De Almar; Enjoy Music with Birute; Elton and Betty White Time; The Jokesters; Lobsterman; The Dollies' Dilemma.

This type of show centers around performing arts, while it is similar to demonstration, in that you are showing the audience something, it is not a teaching show.

For Example

Let's say you are an actor and want to do a scene with a partner. Easy! Make sure you pick a simple scene --not much furniture (the studio doesn't have much) and not too much movement (most studios aren't very big -- maybe twenty feet by twenty feet). The camera people are interns, not professionals, and while some are excellent, others are just beginning, and you'll want to make it easy for them to follow you. The director might not be too skilled either. Network sitcoms use three cameras -- even four and five, depending on the scene. They also use professionals, have a week to get it right and spend a lot more money than $50.

25

How About Some Music?

Enjoy Music With Birtue would feature twelve kids in a circle with violins. Sounds scary? No, not really, since they all played at various times. It was fun to watch them come up and play simple songs -- *Mary Had a Little Lamb,* or *Twinkle Twinkle Little Star.* Then the advanced kids would play classical songs. Birute would have their parents sit in the audience, watch and applaud. It was fun! The kids seemed to like it, and the parents were thrilled.

Want a little spice? Alberto De Almar sat on a stool while performing on his six string acoustic guitar -- he also used an electric guitar. On other songs he'd play along to a percussion sync tape, which filled out the sound. Visually, he first started using a simple background and brought in a flamenco dancer (one eventually started her own dance show). Then he progressed to using a chroma blue screen with special digital computer images flashing behind him. How? Let's talk about some special effects now.

Special Effects

Chroma key allows you to sit (stand, dance, etc.) in front of a blue background, usually a curtain. When I first began working in the studio, we had to use a large sheet of blue paper, ten feet by thirty feet. We would tape it in place against the back wall. It was a nightmare, but we did it. With this blue background you can superimpose any image (still or moving) on it. In Alberto's case, it was moving, computer-generated images from a VHS tape.

For some talk shows, they would superimpose a picture, maybe some artwork, to fill the entire background. The limit to the chroma key was that you could only see the superimposed image on one camera, usually camera two in the middle. When we would cut to the other cameras, you would just see the blue background. Another drawback was that it was sometimes hard to light. You would need strong separation between the person and the blue background. And often it looked like a dark line around the person. However, it was very popular and an effect that people loved to use.

Alberto used a stage that was also covered in blue, so it looked like he was floating as the images raced by. And when we cut to the other cameras (using close-ups only so you didn't see the blue background) we'd use a poster and strobe effect to change the image generated from a TBC. Alberto was the kind of producer who would let you play around and have fun with the effects. His main concern was that you could hear him and see his fingers playing.

Not Interested in Special Effects?

Maybe you just want to do a straight acting performance. Mark Katz had a show that featured seasoned actors performing original one act plays. Similarly, another producer, who had been a network TV director, came in and did a few scenes with professional actors.

The studio became a place where these professionals could work out ideas and scenes without spending thousands of dollars. The only problem was that the camera people weren't used to certain directions and couldn't keep up with the director, who got very frustrated. At times they would have all three cameras on the same actor! Because the interns weren't trained to block their shots, it didn't turn out as will as it could have. (It was actually kind of funny.)

So what the director did was come in with a shot list for each camera person to follow -- just like the network shows. And it worked fine, except when the zoom or focus knob stuck. Then it was chaos again, but in a funny way, as it made every show exciting.

If you're doing this kind of Performance show, you should meet with your director ahead of time to let them know what you need. The studio should know a week in advance who is directing your show. And if the director assigned to the show isn't comfortable with an acting performance, you can request someone else. Ask around -- especially the interns -- they will usually know who is best at directing performance shows.

Always Meet With Your Director

You will need to talk with the director in advance for a performance show. They'll need to prepare and make sure there's enough help. Or you can bring in your own director, if you'd rather.

What About Bands?

Bands? Get your band together and do it. My studio had strict rules about bands playing in the daytime. They thought the noise would disturb the Customer Service reps, who worked right next door. But as it turned out, the reps actually liked it, as the music took their minds off the countless complaints filed by irate customers. But be prepared, you might have to limit yourself to nights -- generally 6:00 p.m. to 8:00 p.m., or later.

I had the good fortune to bring in a band in the afternoon and do an acoustic set, that blew people away. They were called Venice, they were and are a very popular local band, who still play and record. People compare them to The Eagles and Crosby, Stills and Nash because of the way they harmonize.

In contrast, one of the first band performance shows I worked on as an intern, was a heavy metal band that played extremely loud. I couldn't hear the director yelling what to do, so I zoomed all over the place. Later, as a director, I would experience the same problem: the intern camera people couldn't hear *me* over the music. The solution was simple, I would tell the camera people what to do ahead of time. For example, camera three would stay with the drummer and bass player, while camera two would get the singer and guitar player, and camera three got everyone. Sometimes it wouldn't work and the camera people would be on the same person. But we always had to laugh.

Be Flexible

The key is to be flexible and not worry about too many bad shots or funny camera moves. Interns are people like you and me who are learning, and really want to do a good job. Remember to thank them after the show.

Time Reminder

You'll need at least an hour to set up your equipment, at least 45 minutes to tape the show and that leaves you fifteen minutes to clear out before the next show gets there -- unless it's at night and yours is the last show. Then you don't have to worry about another show coming in, but remember your crew. People do want to go home.

What About Your Band?

With three cameras your band will look good, and if you take some tips from me, you can make it even better. Now remember you're using videotape, not film, so it won't look exactly like a professional music video. But, with the right ideas, you can make it look very cool.

We'll say your band has four pieces -- guitar, drums, bass and singer. You can set the drums on a riser in the back. The guitar and bass players to each side, diagonally in front of the drummer and the singer in the middle, just off center -- so you can see the drummer on camera two.

What Kind of "Look" Do You Want? Tips on Lighting Your Band and Other Effects

Now that the band is set up, what look do you want? Here's where you need to make a choice ahead of time. Your background can be a black curtain as a backdrop, which lets you use colored lights to shine on the band members. Ask the director to use two scenes on the light board -- one for color lights, which shine on the musicians, and the other for the white lights, that shine on the musicians from behind (back lights). Also, you'll need a few in front (key lights). This separation of lights will let you vary the look. Without the white lights, the colors will give a cool look to the musicians while creating shadows. But remember, the white lights will wash out (fade) the colors when turned on.

Or you can use multi-colored gels on the gray cyc (back wall) to light it up. The only bad thing is that the white lights tend to wash out the background colors. So what you do is, again set the lights up on two scenes and start with a silhouette, just cyc lights and no white lights, then slowly add the back and key lights, but only enough to get a good level on the waveform monitor (tech talk). This silhouette technique can also be used at the beginning and closing of a talk show, to run your credits over. Also try to set the band up a few feet from the wall, which will allow you to light them and not wash out the background colors. Banners work well too, but can be a hassle to hang. Try to look at the studio in advance for something to hang it from.

On one show I directed, there was a band called The Ironing Maidens -- fronted by three singing housewife-type women in underwear. Although the colors got lost in the background (there were four women musicians behind the three singers) you didn't notice it because of everything that was going on -- three women in front of ironing boards, with irons and spray starch, singing.

Rob, one of the singers' (Kathy) husband, and a music producer (The Eagles, Seahorses) came in to properly mix the sound. He also brought his daughter Abigail Rose to watch her mommy perform. We had a great time! They had the right mix of comedy and musical talent so that it didn't really matter how the studio looked.

How About a Solo Artist?

Another show I directed was just a singer/guitar player named Ken Stacey. I had him sit on a stool in the middle of the studio, and surrounded him with junk -- old lights, trash, a bucket, a mop, a ladder, etc. Then I used yellow gels with a stripped pattern over the lights to give them a rays of sun look, which was layered over red gelled lights. Behind Ken, on the wall, I had bottom lights shinning up with orange colored gels over them and blue gelled lights shinning from above. Then I took the studio monitor and routed one of the camera feeds through it so I could get two images of him at once.

The show started with that silhouette look, and with those background colors it looked awesome.

I had two camera people take their cameras off the tripods and use them "hand-held," which can bring a whole new look to a music performance show. One guy got up on a ladder and got some great high shots looking down on Ken.

32

The trash look came from an earlier show I created called *Glenn's Garage*, where local bands came on to talk about their music.

For another look, a friend of mine had come in to light a female singer/songwriter, Essey, using just white lights and a black curtain. We put her standing up on a small platform, then he put low criss-crossing shinning lights from the floor behind her. But the best effect was when he turned on the smoke machine to cast a cool-looking, almost eerie, fog effect all around her. It settled low and rose in streams through the white lights.

When your budget increases, call around about renting a good fog machine -- remember, make it a good one, the bad ones can choke people and send them running from the studio.

Another type of unplugged or acoustic-only setting can be done with a full band. I did this with a four piece local band, *W*, where I had them set up in a close semi-circle, in front of a black curtain and used the lights -- sometimes flashing -- to make them look colorful. I had noticed that one member had lit some incense and the smoke made a great effect as it rose up to the white back lights.

For another effect, ask to use the TBC (time based corrector), which lets you route a camera or tape deck through it to manipulate the image. A common thing was to use the poster effect to get a colorful, painted look -- many consumer camcorders have this effect. To enhance it, we'd use the chroma to saturate the color and hue to change it. We could also change the bright and dark levels using the other buttons (video level and set up). Then we had the "Strobe" button to make the image delay slightly or longer. The "Sepia" knob would give the image an old time brown look, while the "Mosaic" knob would break the image up into little squares. Although these are used sometimes too much and randomly without any planning, you can be very creative with them.

One trick is to use the poster effect, but pull the chroma out for a black and white look. Or turn up the set-up level while saturating the chroma and use a slight poster and strobe. Experiment and find out the best combinations. A good rule is that too many effects tend to draw attention to the effect itself, and not the performance.

Oh yeah, and as far as sound goes -- good luck. Although they do have a mixing board and some mics, don't rely on them for good sound. I've noticed plenty of bands sound terrible because of the equipment in the studio was not good enough. They are not a recording studio so you should bring as much equipment as possible. Which means your own mics, stands and possibly a mixing board with an FX unit (reverb) and monitors. The guitarist can mic their amp. The bassist can run direct. The drummer can put a mic in the bass drum and two overheads -- one on each side. I used to like putting mics between the snare and hi hat, then another by the floor tom and cymbals. The singer needs to have their own reverb channel. So if you are a musician, you can't rely on the studio.

Where I worked, we had a great mixing board. We had thirty-two channels, a quadraverb unit and some Shure and Peavey mics with boom stands. But the problem was that someone was always fooling around with the mixing board and resetting and re-patching cables. Plus, no one really knew how to mix a band, unless they brought their own sound person. And we could only use eight channels for the mics.

If there is a sound person (intern), and you don't have your own, just record the band for a few minutes and listen back to the tape. You can do this a few times until you get the mix right. Then just have the sound person bring up the master faders.

We did some awesome shows that sounded half way decent too. Some interns were musicians themselves, and you could tell the difference. But, in the end, it all got mixed onto a videotape (not in stereo) and broadcast to a TV -- not a full blown stereo unit.

You Too Can Be a Spice Girl!

How about a Spice Girl type band? You know, the kind that will lip sync a song while doing choreographed dancing. Lip synching isn't a bad word, they do it in every music video. Dick Clark's *American Bandstand* had been doing it for many years. Lip synching is fine, as long as it really is them who recorded the song -- not like Milli Vanilli -- remember them?

The same principles apply here as with a live band, except that you would set the singers up in the studio, play the track through the monitor system, and not worry about the sound.

You might be able to tell by now that I like music and bands. I'm very passionate about working with musicians, and had always gone out of my way to make a show look great. I did a music video once that I showed to a professional music video director, in a class at UCLA, and they loved it.

The video then went on to win a second place award at an AFI/SONY sponsored event. Which then led me to direct (Warner Bros. artist) Randy Crawford's video, *All It Takes is Love*, for an AIDS benefit. I took everything I learned from the studio (working with numerous bands and performers) and applied it to the music video for Randy. I feel very lucky to have had that chance, and the people who asked me to do it were very pleased. The only bad thing is that it was shot on video, not film. Plus, I had to edit it without timecode, in the control room, on older linear equipment (nowadays everything is edited with timecode and on computers). But it was a great experience!

Anyway, let's get out of music and take a look at some other types of performance shows.

Life Beyond Bands

One of the most popular performance shows was *Elton and Betty White.* Here's what the show looked like: Elton, a middle-aged black man and a senior-aged white woman, Betty -- both dressed in bikinis. He played the ukulele and they sang and danced in their own unique way. People loved them because they were so outrageous! They even got voted best local TV show in a local L.A. magazine. People would tell me they couldn't stop watching them because they were so different. But I'd say, they were just being themselves.

I have to tell you a funny Elton and Betty story. When I met them they were guests on another show, and wanted to use an umbrella advertising their phone number, to call them for hire. But where I worked this wasn't allowed, and if I let them do it, my supervisor, who would monitor studio shows in her office, would be down there in a second! (Nowadays I'll see a show where the phone number can be seen throughout it and no one complains.)

Anyway, I told the producer they couldn't do that, and next thing you know they were walking out of the studio and screaming about disrespect. They started getting dressed and threatened to leave. So I booked out of the control room and into the green room to comfort them. Elton was telling me the producer was dissing him and Betty. So I told Elton it's just the way it is and we can work something out. I spent half an hour talking them back into doing the show. And yes, they did use the umbrella. And no, my supervisor wasn't watching at the time. We got away with bending the rules. Phew!

After the show, they asked me about doing their own show and next thing you know, they were starring in their own TV show! I believe they're still on the air doing their thing.

My job had a lot to do with cuddling people's egos and allowing them to feel good about themselves and their shows. I'm proud to say I did it with sincerity, although there were times when I had to bite my tongue and not yell back at somebody. State management and a good quiet closet came in handy -- for screaming!

The Joke's On You

Then there was another show called *The Jokesters,* where Steve and Mariann dressed up like a couple of Jokers (like in a pack of playing cards). Their show was comprised of them performing different on-location practical jokes. They didn't perform much live in the studio, just a few gags between the videotape roll-ins.

What About Magic?

Are you a magician? There are thousands of magicians in the USA, but how many have their own shows? David Copperfield. Penn and Teller. And that masked guy. Locally, a young man named Chris had his own magic show, where he would perform tricks, and show the audience how they could perform simple tricks too. Sometimes he'd even have a guest magician on to do a performance.

Barbra

How about a Barbra Streisand female impersonator? Yes, we had Greg Halstead, who did an amazing impersonation of Barbra. Occasionally he'd do it as a live show (Chapter 19) and speak to the audience while lounging on a couch. He could actually sing like her too.

Greg was one of those multi-talented performers who kept doing a variety of shows. He was already famous for a show called, *The Feelies*, who were

funny masked hosts talking live to people at home.
Later he also had his own talk show too.

So are you going to do a performance show?
Maybe we'll combine everything and do a . . .

VARIETY

Mason Brown Show; Tricks; Not Just Another L.A. Music Show; A Man, a Bass & Box of Stuff; Get the Said Out; Interview with the Batwing; Pamela Roberts Showcase; Cooking with Lenny.

So far we've discussed three types of shows: Talk, Demonstration and Performance. But maybe you're the kind who likes to do a little bit of each? Ah-ha! I know your kind, and that's exactly what this chapter is all about.

Variety simply means any combination of the previous types. David Letterman and Jay Leno have Variety shows -- they perform stand-up monologues, they interview guests, occasionally people come on to demonstrate skills and they have bands perform.

All you need to do is think of an interesting combination. To get your creativity flowing, let's take a look at some simple Variety shows, from where I worked.

Singing Puppets

Probably one of the most popular Variety shows, the *Junior Christian Science Bible Lesson*, was done by David Hart. For his show, he would sit behind a platform and sing Christian songs while operating these puppets: Chip the Black Boy, a Dog, and an Alien (Performance). Plus he'd also have a guest (sometimes an intern or one of our directors) sit in a chair and converse with the puppets (Talk). He simply combined two of the formats. Get it?

Why was he so popular? You could see him performing on the streets, in front of concert halls and Hollywood theaters, singing his songs to anybody who would listen. In a sense he was living his show every day. Once you saw David, you couldn't forget him.

More Talk & Perform

Two other popular variety shows were Ritt Henn's and Mason Brown's.

Ritt, a tall and skinny bass player, did a show called *A Man, a Bass & a Box of Stuff.* He'd interview a guest (Talk) -- sometimes a musician who he'd also perform with (Performance) -- and sometimes do a few comedy skits (more Performance). He, like David, used two of the formats. But he had a completely different show.

Mason was more like the late night guys, as he began his show with a stand-up monologue (Performance), then proceeded to do some comedy skits (more Performance) and even interview a guest or two (Talk). Are you getting the idea?

Totally Outrageous

There was an amazing Variety show that I worked on called *Tricks*. It was a wild dating-dancing-game show. The elaborate set itself would take an hour to set up. Then the show would take about two hours to complete.

But wait a minute Glenn, you said we only had two hours to do the show. Ah-ha! It pays to know the right people, like me, who can show you a trick or two. Read on.

Trick Number 1

They would book the studio at night from 6:00 to 8:00 p.m. -- the department stayed open until 8:30 p.m., and the show before them finished at 5:30 p.m., which gave them a total of three hours -- including that extra hour they needed.

One thing some producers at night would do (when I worked there) was to buy the crew dinner in case they were running late. That way the crew wouldn't mind staying overtime. I can tell you a majority of shows ran until 8:30 p.m., while others could go as late as midnight (they were serving pizza all night!).

Again, I must emphasize that they were all based on the director in charge of the show and what relationship the producer had with the studio. Technically they had to stop at 8:00, but as long as people were having fun and learning, I'd let them stay later (as would other directors).

Another time to book a show was either before lunch (12:30 to 1:30) or immediately afterwards. That would give you an extra hour to go over, or to set up.

Back to Tricks, the Show

They would set up this twelve foot high multi-colored wooden set. We would then pre-tape three exotic dancers, who would perform in a cut-out section, three feet above the floor (to use as roll-ins during the show breaks).

The show itself was set up with fours chairs in front of these structures -- two on each side. The host, David, would stand between the contestants and ask them questions about their dates with each other.

For each correct matching answer, the contestants would get a colorful condom pinned on them. At the end of the show, the contestant with the most correct answers (and condoms) won. Oh yes, and if one of the contestants said a secret word; for example, "love," an exotic (think really exotic) woman would come out with a prize and give it to the contestant, along with a big kiss.

It was a fun show to watch and even more fun to direct. I know it got a few good write ups and reviews (one good one was in L.A.'s *Dramalogue*), but never really caught on with the mainstream audience or networks. I think it was ahead of its time.

Are you thinking of a show like this? I'm telling you about it because I want you to let your imagination run wild and not be limited. It just needs to be planned out right -- that's why I wrote about it and you bought this book.

I'm Too Sexy

Let's look at another big Variety show, that wasn't quite as wild. It was called the *Pamela Roberts Showcase.* Pamela had a modeling business and would use her show to highlight her models. The stage was arranged into a runway, which allowed the models to strut down the catwalk -- just like a real fashion show.

Then Pamela would interview the designer as part of a talk show. And if one of her models had a special talent; like singing, they would get a chance to perform. The main difference was that her show was shot one segment at a time, in order, and primarily didn't use roll-ins. It took a long time to do and seemed to always be at a frantic pace.

Let's Do One

Okay, what about you? Let's take a look at how to put together a Variety show of your own, using the three formats discussed earlier. We'll have a band play music (Performance), a self-defense expert teach you how to defend against an attacker (Demonstration), and you'll interview a movie director (Talk) about their latest movie.

You have two hours to put together a half-hour show. Sound impossible? Nope! First you clear the studio, remove the stage and begin to tape your segments.

That's right, you're going to tape the segments on a separate tape, using a stock tape from the studio, to be your roll-ins for the show.

Light It Up

Lighting will be the key to how fast you can get all this done. Start by lighting the set for your talk show -- using the white lights (key, back and fill) and a simple two-color gel background (cyc). Depending on what you're wearing, you could use blue gels on the sides and purple gels in the middle, or try rose gels on the side and blue gels in the middle. The talk show is the last segment you'll do. Yes I said last. Once the lights are set, you can proceed backwards. Trust me and read on.

Demonstration

Remove the background gels, then just flood (open the barn doors) the key, back and fill lights to evenly light the entire room. Have them boom mic the stage area (or use two wireless lavaliers) as you talk with the Krav Maga self-defense instructor.

Let's say you're learning how to break a choke hold from front and behind. He or she shows you the moves. Then you practice, and actually do it. Cool! Now talk a little more about Krav Maga and you're done that segment. You have an eight-minute segment on a separate tape, with two more segments and twenty-minutes of show time to go.

Performance

Since lighting and set up of the studio, then the self defense demonstration, took forty-five minutes (out of your two hours of studio time), you were smart to book an acoustic act: a guitarist and singer. For the musicians, you can put the stage back in, pull the black curtain and crop the barn doors on the white lights (back, key and fill).

Let's get them on the stage -- put them on stools, not in chairs. I should mention something now about chairs. One of the most awkward things you can do is put a guitarist in a chair to sit next to you. They're not comfortable and it looks kind of silly. The director has to include you (the host) in a few shots, while you stare at the person for the entire song. Unlike the farewell show of *The Larry Sanders Show* -- that's different, people wanted to see him react. Please put the musician in their own space to perform. Thank you, my friend.

What's that? They didn't bring any equipment? No problem, put a lavalier mic on the guitarist and give the singer a hand held mic. Set up time should be about fifteen minutes -- leaving you an hour of studio time left from your original two. Check their levels.

They perform two songs that takes ten minutes. OOPS, they messed up! Add another five minutes, which leaves you with forty-five minutes to finish your show. Remember you'll need real time of twenty-eight minutes to do the show, thank them, get them off and proceed.

Whew, You're Flying Now

Go, go, go! Set the chairs up on the stage, pull the curtain away, replace the background gels, adjust the white lights -- don't forget the diffusion -- put out some plants, get seated, microphone on and you're ready. You should have a good half hour left.

You open the show with the camera on you as you introduce the segments. The director rolls-in the Krav Maga Demonstration. Then you introduce the Band Performance, and the director rolls that one.

Remember the tape time of both segments was about eighteen minutes -- let's add another minute for introductions. Now you have nine minutes to interview the director and then close the show.

Now You Talk

When you come back from the band clip, you introduce the movie director sitting next to you. Wait a minute! How'd she get there? She came on while they were playing the last video segment. They sat her down and miked her up for you.

Give her seven minutes for an interview segment, and use a two minute clip from her movie (make sure you roll the clip in early enough so you have time to talk about it). By the way, did you get free passes to the movie screening?

Suddenly, you're at the twenty-seven-and-a-half minute mark -- you get a cue -- and can close your show.   You thank all the guests: self defense person, band and movie director. Smile and say good-bye. Roll credits over the set and fade to black. Alright, great show!

How'd You Do That?

You might have noticed that the key was the preparation and set up. I've directed Variety shows like this, some more complicated and less organized, yet had them finish on time. It all came down to everybody working together and the host/producer (you) being organized. Do you think you can do it now? Yes? Great!

The Side Stage

I should also mention the possibility of two staging areas. Even though where I worked was considered a small studio, we were able to light a small performance area off to the left side of the main stage. The only problem was that we had limited lights for the area and since it was next to camera one, we could only use two cameras. It was tricky but you could ask for a side staging area for a small band (keyboard player and guitarist) to play during your show and interact with you -- like Letterman and Leno do.

So What Are you Gonna Do?

Review time my friend. Plan it all out. It can be very difficult to pull off, but very rewarding too. Remember you only have a limited amount of studio time. Use it carefully, and review what I've explained.   Start with just two of the types: talk and performance or talk and demonstration. How about cooking and a band? Let your imagination run wild.

And most important, HAVE FUN!

YOU'LL GET IT! ·············▶

JUST YOU

Basic Judaism with Rabbi Mentz; Healing Feelings; It's Your Nickel; Barry Wayne's Venting Event; Stargazing with Anne Shaw; Karen's Restaurant Review; Ask Mr. Traffic; Shirley U Jest; Breaking Through.

This format is popular with singular hosts who mainly want to get personal with the audience -- it can also be the most difficult because you have to talk for twenty-eight minutes and keep the audience interested. Basically, you are trying to connect with someone at home. It can be very inspirational, spiritual, religious or motivating, but also scary if you're not prepared -- don't worry, you have me on your side. Take a deep breath and relax. Good.

A Little Help

Have you ever heard of a TelePrompTer? It is a device that News anchors use to read from -- it puts the words right in front of the camera. If you look close enough you can see their eyes moving slightly back and forth. Most local studios have one for you to use.

TheTelePrompTer where I worked was very old and had to be set next to the camera, so it would look like the host was looking off to the side of the camera when reading the script -- not good. And, it wasn't the most reliable either -- sometimes it would go fast and other times slow. I'd see a host speeding and slowing down according to how it was running that day. It was kind of comical.

But nowadays, I'm sure that most of the studios have upgraded to a consistent TelePrompTer -- better check just in case.

Maybe Some Charts

It's Your Nickel was a show where the host, Barry, gave step-by-step financial advice, while using charts to show you what to do with your money. Although the subject wasn't the most exciting to hear, his hook was that he explained it easily enough for what you needed to do with your money by using charts and graphs. People could visually see what he was talking about and not just listen to him talk about boring numbers.

Think about it. Does your show need something visual to keep people interested, and give them a better idea of what you're saying? A simple graph or picture can go a long way.

Beat a Ticket

Kenny Morse, who has been on national television (Access Hollywood), was the host of a unique traffic theme show, *Ask Mr. Traffic.* His day job was a traffic school instructor. When he did his "live" call-in show, he'd get non-stop calls for the entire twenty-eight minutes!

He sat behind a table -- on it was a small working traffic light and various toy cars -- and proceeded to answer every kind of question about traffic laws. You could see he was actually having fun with the audience as he teased them for breaking the law and then offered advice. He has also won awards for his show, including a Cable Ace. Kenny did a great job with his show being informative and inspiring, while talking about what would be thought of as a dull subject. Who would've thunk it?

How about you? Are you unique in some way that you can inspire people too?

The Ya-hoo Rabbi

I mentioned religious, remember? There isn't any one who could teach you about Judaism as well as Rabbi Mentz could on his show *Basic Judaism.* Even if you weren't Jewish, you could have fun watching him. He had the most energy, with his "Ya-hoo" exclamations! He covered every detail about Judaism -- he even went to a supermarket to tell you if food was really kosher (on location taping is in Chapter 16).

While most single hosts used just one camera, he would use all three cameras. One day we did a show without any interns on the cameras -- it was a beautiful day and they went to the beach -- just the Rabbi and me. It went very well as he would cue me before switching from camera to camera by saying, "come with me." The show's popularity was based solely on his personality.

Do you think you can do that, too?

Hungry?

Karen, another very popular producer/host who had attracted many celebrity fans, did a restaurant review show by herself, with a studio audience. She was fun to watch and listen to -- speaking fifty miles an hour sometimes. Her review would be more like a night in the life of Karen, as she'd tell you what went on before ("getting ready for the limousine to come by"), during the dinner ("how cute the waiter was") and afterwards ("where we went to party").

One show she was talking about Arnold Schwarzenegger's restaurant "Schatzi on Main." Her stories included an event in the bathroom, the model actor-wannbe-waiter and . . . well tune in and watch her, you'll see what I mean.

After she gave you the rundown on the restaurant(s), she would have her little white dog "Clarence" sing. Yes he would sing! Then the camera would pull back, she'd get up from her table and close the show dancing. They would also add a little video strobe effect and a bit of posterization (remember the TBC?), while she danced. There was one show, where a sick child who liked to see her dance, requested a song and Karen danced to it. That was the cool thing about her, she would do the show for more than personal reasons.

Are you thinking about doing a restaurant review show? See how easy it is? All you need is a theme. Just like these people have done. Put together a plan and follow the rest of my instructions.

Relax, Meditate, Get Inspired

Beverly Dennis did her *Healing Feelings* show in a very calm and soothing manner so people could relax while watching. She also had soothing music playing throughout her show.

In the same style were a few other shows. Both Lucy Papillion (*Breaking Through)* and Moira Fox (*On Center)* inspired and reached out to people, too.

The difference was the way they were taped. Beverly let us do these slow camera pans and dissolves all around her and the set -- very artsy. A small wooden figurine carving, plants, lights, her hands -- were all used as focal points while she was talking.

In contrast, Lucy and Moira both kept the cameras on them -- Lucy used a single camera (medium close-up) for the entire show. Moira used all three cameras set (close-up, medium close-up and medium shots). Although different styles, they both kept the audience watching by simply speaking to them. That is talent!

While you're thinking of your theme, think a little bit about your style and technique, too.

54

A Course in Miracles

Marianne Williamson, a best-selling author, who had many celebrity followers, aired a local show too. She wasn't the only one to do a show based on the book *Course in Miracles*. I've worked on at least three "Just You" shows that were based on the popular book.

Is there a book you can base your show on? The *Bible* was very popular with some "Just you" shows. People stood and preached the Gospels for twenty-eight minutes, while others sat calmly and expressed their beliefs.

Need Help?

Suzan Stadner, a counselor, who did a self-help theme show, was also very popular in the Los Angeles area -- celebrities would watch and call her up after her show, or say hello to her while she was out in public. Recently she was on the *E! Entertainment* Channel talking about her former friend, Mrs. Hartman (the wife who allegedly killed her husband, actor Phil).

She'd talk about dysfunction's and abuses that people were experiencing -- relating to them from her own past. One show she used dolls to portray negative types of behavior. You could tell she had been through a lot of pain and suffering herself, but had survived. Now she was helping others who were going through the same kind of pain, and giving them a positive message. She was the kind of person who enjoyed helping people feel good about themselves. Again, she did it all by herself.

Are you the kind of person who can reach out and help someone too? I've been giving you all these examples in the last five chapters, to get you inspired for your show. Need some more inspiration? Okay, how about a little fun?

God & Fun

Antoine had a show called *Kingdom of God,* where he would let the director be as creative as he or she wanted. He'd come into the studio and drape the stage with a gold shiny cloth, surround himself with flowers and Jesus candles.

Keep in mind his attractive stage set up -- you want to catch the audience's eye when they're flipping channels, and that gold cloth sure did the trick!

Once he was set, he'd start speaking about "God" as we'd dissolve in various video images from a video tape of the environment -- lakes, rivers, trees, mountains -- and superimpose images behind him on a blue screen (remember Chroma Key?). One time I got inspired to superimpose Jesus' face (from a candle) on his body -- he loved it! And it fit perfectly for what he was talking about -- "Christ in all of us."

Try It

You can always do a "Just You" show in addition to a talk show. For example, Marketa, who did her *People and Places* show, one day did a solo show, after taping a two person talk show (Monty Hall of *Let's Make a Deal* was her guest). The subject was about her experience as a child in a Nazi concentration camp and trip to a recently completed Holocaust museum in Washington, DC. To complement her dialogue, I would show pictures from her childhood and recent museum trip. It was a very touching and moving show -- one I'll always remember.

Later, when I heard about Steven Spielberg's Shoah Foundation needing volunteers to video tape concentration camp survivors, I signed up immediately and got the chance to videotape three survivors. Again, it was a good human experience. One taping, the person asking questions was "Frenchy" from *Hogan's Heroes,* who later was a guest in our studio.

Do you have a story to tell? Think about it while you read on.

Single Men

Mr. Morrison, a man in his fifties, sat in a darkened studio, spouted poetical philosophy while blowing up long thin balloons. That was it for twenty-eight minutes. And people loved him!

Along the same style, but without the balloons, was Bill Mitchell, who came into a similarly set-up studio and talked politics for the entire show, all by himself, talking -- sometimes loudly -- about his views.

And the audience loved both of these men. Why? Because they were who they were and didn't try to be anybody else but themselves. TV is a place where you can be yourself, no matter what you want to say or do. So do it!

Get Informed

Got a political view? Don't like what congress is doing? Get yourself well-informed and do a show about it. What you need is to know what the heck you're talking about. When you don't, people won't watch. Everyone I've mentioned so far has been very prepared and have made enough mistakes to realize what is working for them. None of the hosts I've mentioned walked into the studio and became instant successes -- all have started with a simple idea and turned it into a TV show that they feel best represents them.

Make sure you are knowledgeable about your subject, and above all, BE YOURSELF!

Meroury In Retrograde

Signe Quinn Taff did her show on-location (Chapter 16) while sitting in front of a home video camera, in her living room, while telling people about "God Centered Astrology." She would tell you what "you" as your astrological sign could expect for the coming months, simply by the way the planets were arranged. Some topics included the "Moon Wobble" and my favorite, "Mercury in Retrograde" -- it's when the machines all start breaking down and chaos ensues (I was actually born during one). She was very serious and attracted a large following.

Again, her show was shot on a simple camera in her living room -- not a studio. There were no fancy camera moves or special effects, just her. It was the message, not the setting that mattered.

Today, you can send her a self-addressed-stamped envelope and she'll send you more information for free. Plus she'll send you information about her seminars in your area -- a recent one in Los Angeles cost $95.00. There was also information about how to buy her video tapes, in case you missed a show.

Make Money

Speaking of selling tapes, when I was working as an intern there was a show where this guy pretended to be this "doctor" character who claimed to be the inventor of AIDS (yes he was serious). I got to be very creative with the lights and I lit him on stage in a very shadowy way that fit the theme perfectly.

A few years later I had seen a video tape in my supervisor's office with this slick cover and picture of him as the doctor. He had been selling them for a profit. The studio's policy was that anyone who made money from the show owed them a professional fee of three hundred dollars for use of the studio.

A few other people had sold shows too. While most studios were fine with just saying, "good luck" and "give us our money," others were demanding more money. If it came down to it, the studio would probably back down and not take the time to pursue you legally -- selling cable is their money maker, not simple local TV shows. But just in case, you might want to ask the studio's policy before you tape there, and read the fine print about their "policies."

But let's not start thinking about money yet. First get your show together. Hey, tired of those other music video channels? Then check this out.

You're a VJ

Another popular "Just You" show was *Dre's World,* which featured soul, rap and R&B music videos. Dre', a former intern who I encouraged to do a music show, would stand in front of a blue screen (Chroma Key again) and project the music videos behind him, then pop in and out.

Ah-ha, now I got your attention! If you look in the back of the book you'll see a list with over twenty music video contacts to call about getting music videos for free! And I mean popular music videos!

I also did a music show, *Pop Bop!* and loved it. Just tune in to those popular channels like, VH1, MTV or BET and you'll see a person introducing videos. You'd probably say, "I can do that better." That's what I said -- and guess what, I picked the music videos, not some executive or committee! And so can you. Let me show you how.

Here's What You Do

So how do you get the music videos? In this book's "Reference Section" you'll find a long list of people from record companies who have sent me thousands of music videos and who will send them to you, too. In fact you'll get so many that you'll be asking them to stop sending them!

60

First step is to send them a fax of what you want to do, and what format you need -- probably 3/4" tapes. Some will reply immediately, while others will need more information about your show and a demo tape (copy of your show). But what if you don't have one?

Just start with the independent promoters; like, Telemotion, Vis-Ability, Aristomedia or MVP, or the smaller record companies like Loud, Mojo, Trauma or Coventry.

Next they'll send you some music videos, which you'll use for your show, and a form to fill out which describes your show, who your audience is, where it will air, what kind of sponsors do you have, do you want CD's to giveaway or tickets to concerts and do you want to interview the bands backstage. One guy's show had free tickets and a limo ride to give away. How? We'll talk more. Fill out the paperwork and send it back.

Say you get the following videos: Reel Big Fish, Cherry Poppin' Daddies, The Mavericks, Blue Rodeo, Everclear (a band I interviewed twice -- once right before they were famous) Megadeth, Dandy Warhols, Shonen Knife, The Flys and Great Big Sea. You now have enough to play for at least one show, if not two. On the average you can get about five videos on per show -- with about a minute to introduce them.

Viewing Them

You can't see them because they're on 3/4" tape and they won't fit in your VCR. No problem! After reading this book -- especially Chapter 7 which explains to you "Where to Go" (a list of studios is located in the Reference Section) -- you'll call the studio to book some time to view them. There shouldn't be a charge since you're getting them together for your studio show. You then pick what ones you like, and also what you think people will like to see, and list them on your rundown sheet (see Chapter 12), one at a time.

Then you do your "Just You" show, where you introduce each clip and maybe even give a little bit of information about the band -- record companies love that, they include bios with the music videos, which tells you about the band and their new album.

Fax

After you do your show and get it on the air, you'll need to fax a playlist of all the videos you've played and when they are airing to the people who sent you videos. This playlist is crucial to them and will be necessary for you to send about once a month. They actually will use your show to help a band's career, and tell people that you are playing it. For me, it feels great to help out a new band!

Variety

At first you might feel obligated to play all the videos, which is fine -- it'll help you mold your show and not limit it. I would play Prodigy, Gloria Estefan, Nine Inch Nails, Dixie Chicks, Morphine, Judy Toy, Reel Big Fish, Cherry Poppin' Daddies, Spice Girls, Madonna, Van Halen, Body Count -- mixing the style so people couldn't stereotype it. It's your show and you can play what you want to. Don't feel pressured to play anyone's video. They need you as much as you need them!

Demo

Now that you have a show on tape, you can send a VHS copy to the bigger companies and get videos from artists like: David Bowie, Barenaked Ladies, Semisonic, The B-52's, Robbie Robertson, Beck, Gloria Estefan, Elton John, Jewel, Sheryl Crow, Sting, Fleetwood Mac, John Fogerty, Duran Duran, Alanis Morrisette, Garth Brooks, Whitney Houston, Brian McKnight, Seal, Aerosmith, etc. (check the Reference Section in the back).

Stuff

These bigger companies will need you to sign a standard licensing agreement in order to get on their mailing lists and receive their videos. You'll also be receiving CD's, singles and albums, T-shirts and stickers, along with the music videos. Stephanie at Virgin sends me everything, along with her best jokes. And when you need to go to a show, just ask for tickets -- sometimes you can get them easily, but other times it might be impossible. When the Beastie Boys sold out The Los Angeles Forum for two nights, there weren't any tickets available for anyone! But if you ask far enough in advance, and it's not a major sellout, more than likely you can get a pair of free tickets to just about any show.

Interviews

If you have the kind of show where you interview people on-location (Chapter 16) you can get an interview easily with the newer bands and even some prominent ones during the sound check. How does that sound? You meet the band, play their video, see their show and listen to their new CD! Plus you star in your own TV show! Life is great!

If you interview a band on location, you can add this "Talk Show" segment to your "Just You" show. All you have to do is edit it together (Chapter 17) and use it as a roll-in for your show.

I'm Elton John and You're Watching Public Access

Well okay, maybe not Elton, but some bands also do video ID's for their videos and your show. I've got a collection of people introducing their videos for my show, *Pop Bop!* Plus I've got some people, like Richard Marx, saying hi to me, my wife and daughter, Zaidee. That is cool! I know my daughter will love it as she gets older and all these people are saying hi to her -- wait until she has her friends over! They'll probably being saying, "who are they?" Guess I better keep doing my show to keep up to date with her musical favorites. Right now it's Spice Girls, Madonna, The Beatles, Shampoo, Sugar Ray, Rugrats and of course, Elmo!

Quick Note

Before we leave this chapter, when you go to the studio with your five to ten music videotapes, have them in order with the rundown sheet. That way they can have them ready for you to introduce. It could be very confusing with all those tapes -- some studios require you to put them on one tape, but since you don't know how to edit (see Chapter 17) quite yet -- you will -- I'm sure they'll understand. Just be organized and know your exact order.

You might even have time to do two shows, but most importantly, have FUN being a VJ starring in your own TV show!

From Your Favorite TV Show

I just thought of another kind of "Just You" show that you could do. There was a producer who was a big fan of the classic TV show, *I Love Lucy*, and what he did was base a show around it. To start, he brought in some VHS copies of the shows and set it up so he could be superimposed (remember Chroma Key?) in the scenes. How?

We would use the blue curtain in the studio and he'd stand in front of it (we'd even cover the floor with blue too). Then we pulled a "Freeze" (on the TBC) from the tape of a scene, without any one in it, from the show, and use it as the background (for example the kitchen or living room). *I Love Lucy* was in black and white, so we'd use another TBC to make him the same color -- I mean, uncolor -- in the studio, by simply turning down the Chroma and adjusting the "Video Level" and "Setup."

It was a challenge, but looked great, and was very unique. He looked like he was on the set, as he talked about the show and gave viewers insights into it.

65

You Inspired?

So what kind of solo show are you going to do? I've
given you a plethora (my favorite word) of examples
to think about. Remember, all you need is a good
topic that you are comfortable talking about and can
do it for a half an hour. Keep in mind the visual
aspect of your show -- an interesting stage set-up,
music videos, etc. Write down some ideas.

Now we'll talk about the place where you make it all
happen. It's called . . .

WHERE TO GO

Just to warn you, this is a long chapter, but very informative. I'll tell you where to go to star in your TV show (I might have mentioned it before) -- a studio, that will cost you either nothing at all (FREE!) or like the studio where I worked, $35.00. That's in comparison to $3,000 a day for a small commercial studio, like the one across the hall from where I worked.

What's a Producer

"Producer" is a word I constantly mention throughout the book. This will be you, as well as the star. So what does a producer do? Producers are organizers. Whether it be a major motion picture or TV show -- they get the talent (actors, guests), directors, writers and behind the scenes people organized. In this case you'll be organizing your own show -- just like we've been talking about so far -- picking a theme, choosing a format and if your show calls for one, booking a guest.

What Kind

You'll also see adjectives before "producer," such as: associate, executive, co, line, etc. They're all referring to a special area of expertise. Executive, is usually the deal maker and money person. Does Aaron Spelling sound familiar? *Dynasty, 90210, Dallas?* Watch any show and you'll see many types of producers. Some, like "line" producers do a lot of work; others, like "associate" are sometimes favors for relatives (Eddie Murphy did this), and it's cool.

How Many

Depending on the show and how complicated it is, will determine how many producers there are. This one movie of the week I worked on had all kinds of producers. The woman who starred in it was one. Hey, let's play trivia: she was in a 1990 movie with Kevin Bacon, Fred Ward and Michael Gross, that I would describe as a kind of old "B-movie monster" type. It was one of my favorites and she was great in it. Can you guess it -- and who she is?

Anyway, her TV movie had producers ranging from her husband, her manager, the writer, the two guys who found the project, to the production company's partners. I think they had about ten producers listed! Plus, her niece had a small role as did her manager's son. The movie won its time slot and made the top ten for the week. Not bad for a country singer (have you guessed her identity yet?). By the way, it cost more than our $50 budget -- more like three million.

And when you get on a big network, you can have as many producers as you want and have all your friends and family on your show. But for now, you as "the" producer will be doing all the work that those other producers were doing (or in some cases, not doing).

Party On

Okay, Producer, let's talk about a secret place you might have heard about in the movie *Wayne's World*, called "Public Access" (also known as "Local Access," "Community Access" or "Cable Access"). The movie and *Saturday Night Live* skit, portrayed Wayne (Mike Meyers) and Garth (Dana Carvey) as two guys with their own TV show, broadcast out of Wayne's basement. In the movie, he had his own cameras and satellite truck to broadcast it -- which would cost a lot more than $50 (you could probably rent a professional mobile studio for anywhere from $10,000 to $1,000,000). Instead, we're going to go to an actual studio to produce and star in your show. (Note: there are studios that have production vans available, but it's not in our budget for now). Later you can explore the possibility of using the van.

Free

Like the title says, you'll be able to do it for $50.00 or less (sometimes FREE). That's right, a Public Access facility will provide a TV studio, three cameras (more on this later), a full crew (camera operators, director, etc.), lights, a dressing room and makeup room. Plus they also have their own channel to air your show -- for FREE. And if you like, you can rent -- or get for free -- equipment to shoot your show while on-location, then return and edit it -- plus they'll train you how to use the equipment for FREE! That is such a beautiful word -- "Free!"

Public Access

No, this is not a late night "No Money Down" infomercial -- it really exists -- probably right in your own home town, too. As I mentioned it's called "Public Access" -- not "Leased Access" -- a commercial channel which we'll talk about later (Chapter 20).

Why do they have them?   It's required by law (the FCC) that cable companies have them -- along with other channels too. Isn't this country great? Without being too boring or technical, the local cable companies are required by law to have a certain amount of access channels -- Educational, Government and Public -- available to the public, when they sign a "Franchise Agreement," with that city (or county).   There was this national "Cable Act" in 1984 that paved the way for these channels. The town where I lived in the Philadelphia suburbs had it for years, but never advertised it to the public. Whereas now, more and more people are becoming aware of it, and after reading this book, many, many more will be using it!

Choices

If you notice you have no choice of what cable company you can choose -- you are stuck with the company in your area. That company, let's say it is Century Cable, made a bid to the city government for your area, similar to the phone company you

have. In Santa Monica we have GTE, in West
Hollywood they have Pacific Bell, but both have
Century Cable for their TV's.

The Cost of Doing Business

These channels are sometimes referred to as "loss
leaders"; in other words, they cost the company
money to run -- say about half a million dollars a
year -- and don't make a profit. Century
Communications -- formerly Century Cable -- where
I worked, had two full-time studios, with about
fifteen staff members. Although they didn't pay very
much of a salary -- $8.00 an hour to start -- they did
have some big expenses, like one light bulb costing
$35.00. If you start to add up those expenses of
maintaining two full-time studios you're talking big
money. Why do it then, you ask? I did mention it's
required by the FCC, but there's another reason
why -- money!

My belief is that if they could get away with not
having them, they would. It makes perfect business
sense to cut your losses, right? If you think about it,
these companies make about $30.00 each month
per subscriber -- which is very conservative.
Century has easily about 100,000 subscribers,
which is $3 million a month gross revenue (not
including all those premium channels and pay-per-
view events, which can run a monthly cable bill up
to and over $100). Say they gross $36 million a
year, and out of that they have to pay $500,00 to
run their required Public Access studio, only
because they have to. Well that's not so bad is it?

It all comes down to the cost of doing business. And also, Public Access sometimes can be used for good publicity and community relations.

Many Access Choices

But you're not limited to the studios you can use. In the Los Angeles area we have Century, Media One (previously Continental), TCI (previously United Artists) and Cox, which all have their own designated areas. Century broadcasts in Santa Monica, West Hollywood, West L.A., Eagle Rock, Silver Lake, Studio city, Sherman Oaks. Media One, is in and around Los Angeles, including Hollywood and Marina Del Rey. TCI and Time Warner covers the Valley. Cox has a little area down past Palo Verdes. (Names and contacts of cable companies appear in the rear Reference Section.) Call up your local cable company for information about their "Access" department. You can also call, in Los Angeles, the L.A. Dept. of Information and Technology (formerly the Dept. of Telecommunications) at (213) 485-2866.

In Los Angeles, the most popular studios with the producers who used them frequently were TCI, in Van Nuys, Media One in Hollywood and Century in Santa Monica, where I worked. They are about half an hour away from each other -- depending on the time of day you drive and the freeway you take.

Noo Yawk

The most popular studio (probably in the nation) has to be Manhattan Neighborhood Network -- they are always being supported by the public, have multiple

72

channels for broadcasting, and are constantly upgrading their equipment. For example, they just bought about a dozen SONY Digital Video cameras -- the three-chip kind (cost about $3500 each) for their producers. That is unheard of in Los Angeles!

The others, BronxNet (Bronx), BCAT(Brooklyn), QPTV (Queens), CTV (Staten Island) and Time Warner (Manhattan), also operate studios. Check the Reference Section for the phone numbers and or addresses. Also, you can call each borough's City Hall for more information.

Back to L.A.

I'll be basing most of what I talk about on Century, where I interned for a year (more about interning in Chapter 18) and worked for 5 1/2 years. Century was also one of the busiest and most popular Public Access studios in the nation. The main reason for its popularity was because it broadcast on Channel 3, which was between CBS (Channel 2) and NBC (Channel 4), and when people turned on their cable boxes it would automatically be on Public Access. This also served for many complaints, too -- more later.

We would also accept tapes from all over the country to air -- there was never a charge for airing, just a simple form to fill out. Just about every state would have a producer sending over a show to air and we'd air it. We got surfing from Hawaii, car repair and motorcycle racing from the midwest, etc. And in return you could get in touch with that producer and ask about their area's policy for airing.

Use The Phone

Okay, Producer, you locate the Public Access studio nearest you, call them up and ask when their "Producer's Orientation" is. At Century it was held generally once month, the first Wednesday of every month at 6:00 p.m. -- sometimes they would have two per month, depending on the need. Get yourself signed up for the meeting. If you feel energetic and have time, call all the studios in your town and sign up for all the orientations -- usually they'll be on different nights. Just don't try to do too much at first.

Mail Goodies

They will send you a package of information called "Guidelines" (explaining procedures and what the facility offers, along with rules and regulations), a sample "Rundown Sheet" and "Credit Sheet" (more on paperwork in Chapter 12), etc. Even if you don't receive this package, don't worry, they will have extras at the meeting.

Congratulations, you've just taken a major step in learning how to star in your own TV show! Keep reading.

Comedy Tonight

Here's what to expect at the meeting. You sit in a room -- could be the cable company's lunch room -- with about forty other people, while a "Public Access Supervisor" informs and entertains you for about ninety minutes. The one at my studio was one of the more entertaining ones -- she had fun stories to tell from her ten years working in the department. People would ask me if we were all comedians at the studio -- we tried to be very entertaining. But if you get a boring person, just stick with it.

Out of these initial forty people, about half will come to the next meeting, half of that will set up a one-on-one meeting and half of those will do a show and only a very few from that original meeting will do a show on a regular (weekly to biweekly) basis.

At the first meeting, listening is the best you can do -- I'll answer just about all of your questions in this book, and tell you little secrets of what to do. Believe me, other people will ask more than enough questions. If you read this entire book before the meeting, you'll know more than anyone else in the room (even some things the supervisor doesn't know) -- better yet, bring the book to the meeting and see if anyone else has read it. Everything in here is generally accepted at all studios.

Next Meeting

After the meeting, you are informed about the next meeting, the "Technical Workshop" -- not all facilities have this workshop, but it is helpful. Usually it was the second Sunday of the month from 10:00 a.m. to 12 noon. At the "Technical Workshop" you are told about the workings of the studio -- what is physically possible (believe me, there wasn't anything we couldn't do, if you can imagine it, we could find a way to do it). Again, for the most part, just listen, I'll explain everything in this book.

75

Also, you'll get to see the studio -- ooooooh. Look and listen!

Another Meeting?

After the technical workshop, you'll schedule a one-on-one meeting with the Supervisor -- who has been entertaining and informing you for the last two meetings. You'll also be told to have your "Rundown" and "Credit" sheets (Chapter 12) filled out for the meeting. Don't worry -- we'll cover it all. I should note again, that not all studios require these meetings, but you should be prepared. Some studios just require one meeting and that's it. I know I said hold off on your questions, but if you hear the "supervisor" talking about something I don't, you should ask a specific question -- might be policy related to how much time you have to do a show, how many shows you can book, how much equipment costs, etc. But more than likely I either cover it, it's in the "Guidelines" or it will be explained at the meeting. Shhhh . . . listen.

New York and other cities might be a little bit different in their procedures. Again, my information is based on the studio where I worked, the studios in the Los Angeles area where I've produced shows from talking to other producers about their experiences and reading studio's guidlines.

Okay, time for your one-on-one meeting -- let's say it's at 1:00 Wednesday afternoon. You work during the day, and they only have weekday meeting hours, so you go during your lunch.

Did you read over your package? Did you fill in your rundown and credit sheets? Are you ready to talk enthusiastically about your show, for an hour? You must be realistic about what you want and need. This book will prepare you for every question they can ask you about your show. So read on.

The Supervisor

The supervisor you're meeting with might like to talk and talk. Usually, these meetings are the majority of what department supervisors do -- along with some paperwork, meeting with the cable bosses, gossiping on the phone, balancing their checkbook, sneaking off to Disneyland -- yes, sometimes they need to get away from the office. They rarely direct shows, just manage the department (which sometimes can be difficult) and therefore aren't as familiar with the actual producing of the shows as the other full-time directors.

Just the Facts

When you meet with him or her, you'll go over your show. Don't be too revealing and get into lengthy conversations, just give them enough information to let them know you're sure of yourself and know what you're doing. Let them ask you questions -- volunteering information could drag the meeting on for too long. Your goal is to get in and out in thirty minutes.

My supervisor would average about an hour per meeting. I would get a meeting done in about fifteen minutes -- not that I was any better, just that I knew, as a director, what I needed. That would mean all a producer had to do was assure me they knew what they were doing. And most people did.

If they didn't, I'd tell them very simply how to do it. Just like I'm doing in this book.

You have a "Rundown" and "Credit" sheet, and that's all they need to see. It tells them everything they need to know for your show. If a producer didn't have one, just as long as they came in ready to go and excited about their show, I'd tell them to work on one and be prepared -- also to call the director a week before their studio time, which is very valuable (probably more than meeting the supervisor). This way you can go over your show a week ahead of time with the actual director and he or she will tell you what you need. See how easy it is?

Studio Time

Now you book your studio time with the supervisor (again some places are different and do it over the phone). Where I worked it was in two hour blocks Monday through Friday: 8:30 a.m. to 10:30, 10:30 a.m. to 12:30 p.m., 1:30 p.m. to 3:30 p.m., 3:30 p.m. to 5:30 p.m. or 6:00 p.m. to 8:00 p.m. We had lunch from 12:30 p.m. to 1:30 p.m. On Saturdays, we'd have 9:30 a.m. to 11:30 a.m., then the "live" shows 12:00 noon, 2:00 p.m. and 3:00 p.m., where you'd have half an hour to set up and half an hour to do the show (more on live shows in Chapter 19). Choose a time that is realistic for you -- a time that will allow you an hour before and after -- figure four hours of time for your show. You might have to wait a month or two -- depending on your needs and their availability -- in some cases you can get in sooner.

Can I Really Do This?

I should take a moment to warn you about what you can and can not have on Public Access. It is probably best to double check with the supervisor before doing something you're not sure about. Again it should all be explained in the guidelines, and I'll cover most of it too.

Sponsor No No's

Okay. Commercials. You can't have a sponsors' address or phone number, or give out how much they charge for services -- and they prefer you not mention the sponsor during the show or show their logo or product. If you are cutting hair, don't show the name brand of a special shampoo or spray and say it's the best on the market. Do it subtlety -- "I use Joico jojoba shampoo, but you can use whatever you like." Get it?

There you go, you just mentioned a product, and no one can say you were being "commercial."
One of the studio's rules was, if you mention a restaurant, you had to mention two others, the same with a book or any other product you liked. I did a show about Gibson guitars once and they were upset with me because the guy was raving about Gibson guitars -- we had ten in the studio and he was jamming -- it was great! Eventually I did get to air it and no one got upset. When I reviewed restaurants on *L.A. Flavor*, we would make sure to mention two other restaurants at the end of the show -- saying just their name and city.

Just the other night there was this different perspective type show (the host was a former intern that I knew) about a personal workshop course, and the guest and host were raving about how it was the best and what you could learn. And guess what. The studio who was always complaining about commercial shows, was where it was produced and they aired it! Ironic? Just be careful with commercialism -- it can prevent your show from airing.

Your Birthday Suit

Nudity. This subject always came up and it was a major cause of headaches for Public Access supervisors and us directors. Here's what the studios would tell the producers, "No nudity allowed." The New York studios were more liberal and, I've heard, that you could show full frontal nudity. But here in Los Angeles, they wouldn't allow it at all -- up until the last few years when they started allowing some "nudity," but only if they were scheduled late at night

I've heard the word "obscene" (no medical value) spoken a lot around the studio, and explained like this, "a breast exam is fine -- but don't show breasts, a man's penis, or a vagina in any other way."

First Amendment

Here's what nudity you can really get away with: ANYTHING!

80

That's right you can show anything you want to, because it's your first amendment right! And every time our Cable company was challenged on this issue, they backed down! Why? Because you have to pay for basic cable -- "it is something you have to ask for" (that was the standard answer). It is not free, and therefore not regulated by the same government agencies like the FCC.

Do you wonder why radio shock jocks like Howard Stern or Don Imus can't say those curse words on the radio? It's "free" broadcast and controlled by the FCC. The cable studio can argue that people have to subscribe and ask for cable TV; therefore, should expect to receive some unregulated programming -- if you ask me that is a cop out!

Mistress Julie

Here's a good example. Julie, the host of the *Mistress Julie* show, was in my "Remote Camera/Editing" class, and she asked me about taping a penis piercing. After I cringed, I told her I couldn't tell her what she could or could not air -- I could suggest to her that I didn't think it was appropriate, especially for young viewers, or I didn't care to watch it. My response was all from the "I" viewpoint. From "her" viewpoint this show was fine and her audience would love it. Instead of trying to convince her of my viewpoint, I assured her she could do whatever she wanted to do. She was surprised! Then she went ahead and aired her show that featured a penis piercing, at 9:00 p.m. This wasn't a good time to air it -- the policy was that it "should" be after 11:00 p.m. Oh well.

Uh-oh

This show with it's very "graphic demonstration," drew a lot of complaints from viewers. But they were told the standard reply, "Turn it off, don't watch it, there's nothing we can do, write to the producer, etc." My supervisor complained to Julie about her show. Of course Julie said that I told her she could air it -- which was correct. Basically, Julie got a lawyer, the press got involved, a big deal was made about being censored, and the company looked foolish! One of the executives went on the news and gave them the same explanation as I just did, "It's their first amendment right. . . ." Good for Julie!

Remember Colin?

Another example was *Colin's Sleazy Friends*, where one of his guests stripped naked at the end of the show and started dancing around -- the camera people loved it as they zoomed in . . .well you get the idea. Again, "nothing we can do," was the answer given to irate callers. He did his show in our studio from time to time -- I directed his first one when he didn't have guests (or nudity) and was just starting out. He had some roll-ins from an adult movie convention in Las Vegas -- he was actually a funny guy!

I'd always give the interns, male and female, a choice of working on the show or not -- some didn't want to and left. He could be very entertaining to watch, along with Dino his co-host. But eventually I chose not to work on his show anymore. We always got along, but as my life was changing for the better, I felt I couldn't be passionate about directing his show. And believe me, there were plenty of other directors who jumped right in there to work on it!

The Doctor Is Home

Some more sexual based shows were *Dr. Susan Block*, who, although wasn't nude, played with sexual aides while rolling around in bed, dressed in her underwear, and giving out "advice." Her husband taped the show out of their home, using a very stylized effect -- the camera would pan around while they used a slight strobe effect. It eventually got picked up by HBO.

Along with Colin, she was probably the most popular host on Public Access. If you're thinking of doing a show like this -- that can't be shown on network TV, be careful of policies and regulations, and be ready to stand up for your rights!

A High Priestess?

One of the strangest shows was a lady who called herself the "high priestess of love." She supposedly had a religion based on sex, where all members had to have sex with her. She would show nudity on just about every show -- even her own sometimes, which as I was told, "was pretty disturbing." Eventually, I understand, she went to jail for really running a prostitution ring!

7 Deadly Words

Curse words were even less threatening. One music video show was out to desensitize the audience to the infamous "F" word. That's what they actually told my supervisor when she confronted them about excess usage of the "F" word. Then they called the L.A. County Supervisor, (213-485-2866) and complained that they were being picked on. And guess what? They got their way and continued doing what they wanted!

Strong Stomach

Caution: what I now describe could make you feel ill. We used to screen tapes that weren't produced in our studio. The majority were fine. But there was this series (I won't say the name to give it anymore publicity) of shows that came in that just disgusted me. It had no artistic value what so ever. Some segments featured a naked singer (I should say screamer) who defecated in a crowd and then ate it. Then the host in another segment pulled out his penis and stuck it in a plastic doll. (Sorry, Reader!)

When I saw shows like that, I didn't get what the people were trying to accomplish. What kind of audience will tune into that? If you're reading this book, I'm sure you agree with me -- it's trash! I'm not on a crusade, or trying to tell people what is art and what is not, but c'mon people let's be honest -- there's something just not right with that kind of TV show. I'm just sorry I had to see those images.

My Soapbox

I don't condone these kind of shows that flaunt nudity, obscenities or use excessive curse words or destructive behavior, because they are a person's point of view. Personally I don't watch them, and I'd prefer they not be on when young kids could be watching. However, I do believe, as a parent, it is a parent's responsibility to raise their children functionally, and not rely on the media to provide good values.

What kind of examples and values are you going to give to your audience? Make sure you believe in your show and what it stands for -- you might have to defend it someday against a moral majority.

Problems?

If you run into a problem with a studio, then you can call the department that oversees all county Public Access studios. In Los Angeles, that would be the Bureau of Information and Technology (213) 485-2866 (or call the Mayor's office and City Council, they always like to hear from the voters -- especially during an election year).

Be Afraid

Cable companies are afraid of the "County Supervisors," who oversee all of Public Access. Why? Because they have the power to impose large fines on cable companies. There were times when we had heard this supervisor "Tony" was coming by for an inspection, and our supervisor would panic. If Tony didn't see everything working properly he'd make "recommendations." Which meant fix it, change it or face a fine!

Tony was actually a very nice man who had previously worked in our studio -- long before I came along.   But he had to make sure the studios had everything they were required to have for the public (you) -- he was good to know.

Back to the Package

Okay, now that I covered the "No no's," let's move on to some paperwork they gave you. Somewhere in that package they gave you is an area called "Rules and Regulations." If not, I'll tell you what they are. We already touched on obscenity, nudity and commercialism, but now there are some hard facts about what you can have. Remember these are all based on Century's rules and they vary according to the studios and the city's where they are located -- but generally all are the same.

Studio Time

You can only have two hours of studio time on the books. The books go for an entire year, so that's all you get. After you tape a show, then you can book more time (probably about six to eight weeks later). If you're doing one show per hour, you can book two separate hours on two separate days, but still only have two hours -- only a handful of experienced producers did it this way. Those two hours mean you'll probably only tape a show once every six to eight weeks in that studio -- longer if you're doing it at night.

86

I Get Around

So how do you tape a show on a regular basis?
Easy. There are about five studios within driving
distance, in Los Angeles -- Century has two that are
about forty-five minutes apart -- you can have four
hours divided between the two different studios. TCI
is in the valley, Cox is thirty minutes from there.
Then there's Media One in Hollywood and the
Marina. One week you book at one, the next you
book at another and so on. And you can keep this
going as long as you come up with topics and
guests -- usually some guests are closer to one
studio, too. It gets to be like working out and
exercising -- like an addiction that is good for you!

A Little Secret

How else? Here's a little secret, it's called
"cancellations." That means when a producer can't
make a scheduled taping, they cancel their studio
time -- at least twenty-four hours in advance -- or
they're supposed to be charged (we never charged
them -- we enjoyed getting a break).

Most producers will cancel ahead of time -- two to
three days, and that means you can get more taping
time, at a studio close to you. But you must be
ready within a week's notice, sometimes a day.
You still get to keep your original time too. For
example, you call up on Monday and they have a
cancellation that Thursday, you can still keep your
other two hours the following week. If the
cancellation is more than a week away, they might
cancel your other booking -- it depends on who's
answering the phone and who's checking the books.

Oops

On several occasions, my supervisor had me go through the studio book and find all the people who had more than two hours booked. I found about a dozen who had more time, plus there was one lady who had ten hours booked. How'd she do it? Her friend (who eventually got fired) was in charge of the studio book. When I found people with too much time, I had to call them up and tell them to cancel a booking -- some didn't even know they had that much time booked.

Ironic?

But the funniest part about Public Access was that it was the producers who were complaining about each other getting more time. They would turn each other in, and the guy in charge of booking time was doing all these favors for his friends -- it was a mess. They would actually tell me about what another producer was doing wrong, when I would call them up and tell them that had to lose some of that overbooked studio time. Then they'd complain to my supervisor that I was picking on them -- she had a soft spot for a few of her friends and let some of them keep their time.

A friendly favor example was during October 1996 -- a staff member's roommate had four "live" shows in a row -- they were all "cancellations." On a regular basis you could get one to two "live" shows about every six months. But because they lived together, he had first pick at what was available. When the staff member saw the opening she booked him (or had one of us do it) and sometimes listed herself as the producer -- but was never there for the show.

Every other staff member would do favors for friends too. It was a very mis-managed situation.

There were times when people would call in for cancellations, and the person answering the phone would be staring at a cancellation and say, "Sorry I don't see any right now, try later in the week." Simply because they didn't know them, didn't like them, were too lazy to do it or was holding it for a friend.

Cleaning Up

I did hear that they got the Local Origination manager, John, to oversee Century's Public Access departments and it is now easier to get a cancellation. He's a good guy -- we worked together, so mention my name. If you don't get satisfaction from the department supervisor, always go to their boss -- in this case John -- (310) 315-4400.

When I worked there, Bill Rosendahl (currently a Senior Vice President) was the man to call for fixing problems quickly. We produced a show together for Kelly and Sharon Stone's homeless organization, Planet Hope. If he could, he'd help everybody, but eventually he got very busy with his own show, *Week in Review*. Now John oversees the departments. Be nice when you call these people, explain exactly why you feel you have a problem, and know what the rules are.

The reason I mention these behind the scenes episodes to you (believe me there are a lot more that I could tell you about) is because it gives you a understanding of the people you'll be dealing with.

People are human -- don't' expect them not to make mistakes or be swayed by an emotion.

Food and Roaches

Okay, we got your studio time and you know how to find a cancellation. Let's talk about food. United Artists, at one time had a policy where they insisted that you bring food for the crew. What they would do is post your show up on a board and the interns would sign up to work on various shows -- depending how good the food was. There was one guy who had it catered every time! And he had the most interns working on it!

But this rule eventually got thrown out. The food would sometimes lay around and start to smell, then the roaches would come out. Eventually the policy changed to no food at anytime! Ask the studio directors what their policy is (written and unwritten) towards food and getting a crew to work on your show.

Pizza Time

At Century, you would be limited to an area where you could have food -- the green room only. But we would bend the rules so that the interns could get fed -- they were neat and cleaned up. If we were running over at night or lunch, because the producer wasn't prepared, we'd ask them to buy the crew pizza. So if you were booked at 6:00 p.m. and ran past 8:00, you'd order a pizza or two, and could stay an extra hour -- sometimes until midnight. Yes it did happen. It was a barter type system that worked well for everyone.

Technically we could shut a show down or charge them double for each half hour they ran over. Pizza was cheaper. Again, this all happened when I worked there -- it might have changed.

No Alcohol

Before we end the chapter, there was a policy about no smoking or alcohol on the premises. But we had a regular beer-making show, where the people were slowly getting drunk in the studio, on TV.

Remember, there are always exceptions to rules. Not that you should think about bending them or breaking them, just know what the people at the studio will allow you to do. It might also take you a little while to get to know the staff and for them to know you, but you can have a great relationship with a studio, and that will make your show even better.

Summary

Let's make sure we know where we are. From this chapter, you've learned about "Public Access," what it takes to be a producer (meetings), and you have your studio taping time booked (in the future you can book right over the phone) at a Public Access Studio -- maybe even two or three different studios. We also talked about some policies and "no no's," and the "Guidelines" you need to read. Plus you have your theme and format, and maybe even a guest or two.

Now you need to prepare for the "Big Day" -- that's Chapter 14 (we have Chapters 8 to 13 still to read)

-- which may be a month or two away. Cool, it gives you plenty of time, but don't waste it. You have more "fun" work to do (and don't forget the pizza money -- just kidding!).

But first, let's read some more chapters -- starting with getting some supplies together.

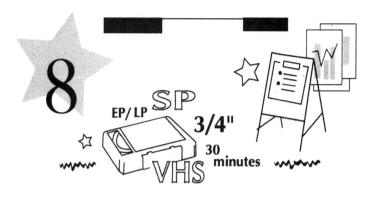

SUPPLIES

Now somewhere in the meeting or in the paperwork you recall something about a 3/4" tape. What is a 3/4" tape? Okay, let's talk about that and other stuff you'll need.

You'll need a thirty minute 3/4" tape, your broadcast master, to record your show on, so they can air it. You can buy one at Studio Film and Tape, -- get an "Eco Tape" (previously used and recycled) for about $6.00, or a new one for about $12.00. Brands don't matter much, SONY BRS 30 is good, Maxell KCA 30 -- you don't need the more expensive SP, because most Public Access channels don't broadcast them (check just in case).

Also, check if there are any discounts for being a Public Access producer. Studio Film & Tape offers 10%. The president, Carole Dean, has been doing TV shows for years -- *Healthstyles* and *Filmmakers*. She's a very nice lady and big supporter of Public Access and independent projects. That's why I recommend her store, in Los

Angeles, New York and Chicago -- see the Reference Section for phone numbers and addresses. If you don't go to her store, then call around town -- look in the yellow pages under "video supplies" for stores, or ask the studio where you can get tapes.

VHS Tape Too

You'll also need a thirty-minute VHS tape, about $1.00 at the same store. Or buy one of the regular tapes at your local supermarket -- usually you can get those two to six hour tapes in a package deal. The studio decks record at SP speed, so you can get four half hour shows on one tape; however, the problem is that if they don't cue your already-taped-on VHS tape correctly, they could tape over a show. I strongly recommend using one half-hour VHS tape per show, that way you won't have to search for your show.

Sp, Sp, Sp

Speaking of the studio VHS decks, if you have a roll-in on VHS tape that you want to use during your show, make sure it was recorded at SP speed. Many times a producer would come in with a great segment, recorded at EP or LP speed, and we couldn't use it. Also, the studio might have only one VHS deck for recording or playback. Again if your roll-in is on VHS, then you'll have to make a VHS copy of your show later. Or better yet, have the director bounce your footage up to a stock 3/4" tape and use that as your roll-in. That is, if time allows for it -- it should if you plan carefully, and listen to me.

You can later buy a case of 3/4" and VHS tapes -- they are cheaper when you buy ten, but wait until you do a few shows and have a bigger budget. Again, Studio Film & Tape (323-466-8101) has them.

Stock Tapes

If you don't want to spend the money on a 3/4" tape, the studio should have a stock tape you can use to tape your show on. It is also a recycled tape, but no telling how many times it's been taped over. They will air the show and then tape over it, or you can buy it, but it costs about $15.00. I remember giving them away all the time, we had so many of them it didn't matter, and people really appreciated the gesture. Also be careful not to rely on the studio for a 3/4" tape -- if they don't have a used one, you might have to buy a new one -- which could cost you double the price.

Funny Story

We had a show from Palisades High School that was recorded on a "stock tape." The original tape was donated by the *Roseanne* show. Well, guess what? One time the school show was short, and the end of an unaired and unedited *Roseanne* episode played on Public Access. Of course, ABC called the station immediately, but still too late to pull it off the air.

Wow! What A Quick Chapter

Okay, you've got your tapes, a 3/4" and a VHS tape.
What else could you need? Basically that's it.
You'll walk in with blank tapes and walk out with a
show on each.

As far as the stage goes, they have everything you
need for basic shows -- chairs, plants, tables, etc.
As you get going you can bring in some of your own
furniture, plants and props for the stage.

Now, let's talk about how to get some great music
and maybe even a guest for your show.

STAY PLUGGED IN!

MUSIC

Every show has a theme. Remember *The Tonight Show? Entertainment Tonight* has that catchy (or annoying) theme. And just about any sitcom or drama has or had a memorable song (or at least they hope so). Think about *Batman, The Jeffersons, Good Times, What's Happening, Fresh Prince of Bel-Air, Miami Vice, Cheers, Taxi, MASH, Buffy, Dawson's Creek, Friends, Homicide, King of the Hill* and *The Simpsons*.

Elton's Song?

Before you go thinking of using the popular songs by The Beatles, The Police, Elton John, Billy Joel, Fleetwood Mac, Bruce Springsteen, Joan Jett, Carly Simon, The Ramones, Pearl Jam, Beck, Garth Brooks, Travis Tritt, Puff Daddy, WuTang Clan, Reba McEntire (remember her in the movie *Tremors*?), or some defunct band from the 80's (plenty of those), remember we only have a budget of $50, and that wouldn't cover the cost of using as little as 10 seconds of one of their songs.

Big Bucks

The reason? There are two big Performing Rights organizations, BMI (which I am a member) and ASCAP, that collect money for their songwriters and bands -- these amounts are huge. Whenever you hear a song on the radio, on a jukebox, in a club, on TV, at a ball game or anywhere publicly, you can bet someone is paying money for that right, and BMI or ASCAP is collecting it.

Royalty Free

So what do you do? Read on my friend. There are special CD's you can buy that are royalty free -- which means you don't have to pay a fee every time the song is played; however, they can cost two to three hundred dollars to buy. Over the course of twenty or thirty shows, it'll pay off, but that is not in our immediate budget. Also, those CD's are very sterile and generic ("cheesy") -- especially for your theme.

Look Around

Earlier I talked about guests for a music video show and recommended a local musician. There are many musicians out there that will donate music for your show in exchange for a simple credit: "Music by _____." All you have to do is ask them. And all you need is one yes.

Please Release

You will want to get a "Release" form for them to sign. In it, make sure that they will let you use the song for your show without any stipulations -- free and clear for the life of the show. The studio where you are taping should have a "Release" form available for you -- make a lot of copies of the blank one. If it doesn't say what you want it to say, then re-write it on your computer to be directly related to your show and needs. Also make sure you give them a copy of it and understand what you will do with it. You might have to negotiate a little here and there, or if they are waffling, forget their song. This happened to my wife and I with a song. We were ready to drop it from a program -- they wanted a limit on how long we could use it and wanted to re-negotiate every so often, which is great for them, but in our case it wasn't feasible. So eventually the person came around, and it turned out everyone was happy.

Venti-Soy-Extra-Hot-No-Foam-Latte

(That's my wife's drink.) Say you hear this duet in a local coffee house -- an excellent place to go -- and like their music. After they're done you can approach them, and ask them for a cassette -- which they should have available for sale or promotion. Since you're on TV and can give them exposure, ask if they will give it to you for free.

Hi, My Name is Glenn and I Have a TV Show

Tell them about your show and that you're looking for a theme. You can even offer to buy them a coffee drink. If they like what your show is about, they'll give you a cassette, or send you one. Listen and find a good song, then call them up.

If they let you use a song, get the release form ready and mail it off. You'll also want to keep them up to date about how your show is going, and you might have them as a guest on a future show -- a nice return gesture for using their music. It's great how this deal works -- musicians get exposure and you get a theme.

The R.E.M. Experience

If you're still bent on getting a popular song for your show, let me explain what I did. In 1997, I made a half-hour documentary/show about a non-profit organization called Planet Hope -- they help homeless families get jobs and back into society (founded by Kelly and Sharon Stone). For the theme, I wanted to use R.E.M.'s *Everybody Hurts*. On the back of their CD, I found the record label, Warner Bros., and the publishing company, Warner Chapell.

I called Warner Bros. Records in Burbank, got in touch with the licensing department -- they own the rights to the recording of the song -- and listened to the instructions on what to do. I sent a letter of what I needed -- permission to use the entire song in the show, describing exactly what the show was (documentary) and where it would be shown (Public

Access in Los Angeles) and for how long (six months). They needed all the details.

Within two weeks I secured the broadcast recording rights for free, from Warner Bros. Records. Planet Hope being a non-profit organization certainly did help, as did Public Access being a non-commercial channel.

With that momentum, I called the publisher, Warner Chapell in Los Angeles, got a contact name and then sent another letter, explaining that I had the rights from Warner Bros. Records, and gave all the details again. Within two weeks a woman called me back and we talked about it -- actually I pleaded nicely and negotiated the price of $100 for six months usage. But she told me she would have to get in touch with the band to get their permission first. Two weeks later I had a letter/contract giving me permission to use the song and the credit they required. I sent in my $100 check and started editing the show.

There Are Others

Some lesser known bands will give you permission directly, if they like what your show is about. Again, it helps that Public Access is a non-commercial channel -- you're not making any money. And if a person likes what you're doing, then they will give you permission. Still, there are even obscure musicians who want money for their music -- choose carefully, and don't forget that release form.

Take a Chance

Lastly, I'd like to let you know that if you do use a song -- for example a woman had been doing a show for years, using Sting's music -- I believe it was called *Deeper Wave* -- and never got in trouble. She did give credit at the end of the show, and no one said a word. So yes, you could get away with using someone else's music, but when you get more popular, then you are definitely going to get caught and have to pay. And you know, it's only right that you pay someone for their music (or at least get their approval), they did have to create it, didn't they? As a musician myself, I want to have say about my music too. Also the publishing and performing rights companies are taking a closer look at Public Access -- be warned!

Public Access Library

Your last choice for music, but eeriest to get and least rewarding, is the CD selections they have at the studio. During my last year, they invested in some of those "Royalty Free" CD's, which sound a little better than the elevator music from the 70's . But they were used by a lot of other shows and had that "sound-alike" quality of all the rest. However, they are there for you to use, and can come in handy when you don't have time to look or shop around. The studio was paying a yearly fee for using them, but you might run into a problem later if you try to sell your show.

Also it'll take time to listen to them and that takes away from your two hours of studio taping time. Best to have your theme ahead of time.

Changes

Remember you can always change your theme later -- when you really get going! I've changed mine two or three times to update it and keep it fresh. Don't feel locked into a theme if it doesn't work for you anymore. But also, if your theme is catchy and people remember it -- keep using it.

Let's Get a Cup of Something

Okay, you're now in search for music. Do you want to take a break while you enjoy some music at a local coffee house? Bring the book along! People will be curious as to what you're reading and what kind of show you're doing. You might even find an interesting guest.

YOU'RE BARROWING
RIGHT ALONG!

10

SPONSORS

Nike or Converse? Coke or Pepsi? Bud or Miller?
Mercedes Benz or Ford? So what do they have in
common? They pay big bucks to people to
advertise their products. What can they do for you?
Nothing! That's right, nothing!

How About a Limo Instead?

Why? Because you're on Public Access and they
don't allow commercial sponsors. Well how about a
free haircut, free airfare, free hotel accommodations, free limo usage or free flowers? Forget soda,
that costs less than a dollar. You know how much a
limo costs to rent?

There are a few shows that get limousines to drive
them around, for free. James has his *L.A. Nitelife*
show where he's usually cruising the town in a limo
-- for free! Earlier I mentioned Karen's restaurant
show -- she also gets limo service to and from the
restaurant -- just like a real celebrity, well guess
what, she is (on a local level). And after I'm done

with you and you get your show going, yes, you too, will become somewhat of a local celebrity -- not for celebrity's sake, but because you're starring in your own TV show.

Limos are a blast to cruise in. I shot a show once where it was just me with my camera and the producer talking about the show. As silly as the show was, it was fun to be in the limo, and I got a different perspective on the world and how it feels to temporarily cruise in the lap of luxury. Cool.

Free Haircut

How do you do it? Easy. Let's start with your first show. You need your haircut and you have been going to the same person for years. You tell them about your show and how you would like to give them credit in exchange for a free haircut. They hear the words "TV credit" and you have your free haircut. Yes, it is that easy. Just try it. And you can even thank them verbally at the end of the show, ". . . and a special thanks to Gib for keeping me looking so good." They'll be thrilled!

So why would your hair stylist want to have their name on your show? It's TV, and getting credit is a big deal. And if someone watching likes your hair style, they call you up and ask you how to get in touch with your stylist. Now they have another customer, and they tell a friend and so on.

Warning Warning Will Robinson

Public Access channels are cautious about sponsors and commercialism -- we spoke about it briefly and I'll cover it again. Basically, you can give a sponsor's name and city only, for ten seconds as a credit. On your credit sheet (Chapter 12), you write "Hair by Gib of Santa Monica, CA," and they'll run it at the end of your show. That credit is not commercial since you don't have their phone number listed, or what they charge for their services, and that is the key to sponsors. Again, this is based on where I worked -- check with the studio just in case, but they all operate basically the same.

Too Many?

I saw a woman's show that had over fifty sponsors, from dry-cleaning to logo design to flowers -- it took two to three minutes to run them, and each one wasn't on for that long -- maybe two or three seconds. And from what I hear, she got some free services from all those people. We'll be limiting our show to one, two or three sponsors, that way each one will feel like they are really contributing to your show, instead of one amongst many.

Warning . . .

I'm beginning to sound like the "Robot" from TV's *Lost in Space*. Some Public Access channels limit you to three sponsors. Always check with them ahead of time. Hey, it's a question you can ask at your "Producer's Orientation" meeting. If you see a show with more than the allowed sponsors, write it down, and if they tell you, you can't have that many, ask them why another show can. That's how you have to do things sometimes. It's only fair that if someone else can do it, the rest of us should be allowed too!

Timing

Remember the limo sponsors? The first company you call might not be so generous. It's all about finding the right people who like what you're doing. Try the person's show you saw a sponsor on and ask if they would like to sponsor you too. Again you could try the friend of a friend of a friend approach -- it's always better to have a name of a person who a sponsor is familiar with. If you don't, you have to keep asking until you get the right response -- I know, I hate rejection too -- do you know what I went through in writing and getting this book published? -- Whoa!

Watch Out

I remember one show, where the producer got a friend to hook them up with free airfare to a resort and free hotel accommodations -- including meals -- but later got turned down for airing because it was too commercial for most Public Access channels. The problem was that they raved about one particular place too much and it turned into an infomercial. And when you can't deliver your air time, then you've got a serious problem. That's why we're starting small and going over all the rules.

It's All in the Wording

There's an art to not being commercial. It is all in the wording of the show. No logos, no over hyping the places you go to, and one key is to compare and offer other selections. If you have someone talking about a restaurant you should mention two other restaurants that are just as good. But remember, no prices and don't say how they're the best in town. Be cool, or you get no air time my friend, and that means the place you went to is going to be upset that they're not on the air! When you graduate to the commercial channel, you can advertise your heart out!

Connie Knows

Connie Martinson, on her book show, takes time at the end, after speaking to the author (sometimes two) of his or her book, to tell viewers about a few other books and to "go to your local library to read the book." Yes, she talked the entire show about just one book -- almost like an infomercial in disguise. But she really was (and still is) encouraging reading -- you could tell by her passion.

Big Brother?

However, when I would direct her show, I was warned by my supervisor not to show the book too much, since that was supposedly "commercial in itself." I'd have the guest sit with the book close to their shoulder so you could see it in their medium shot. And wouldn't you know my supervisor would be watching the show in her office, then call down to the control room for me to tighten up the shot (this was the same person bending the rules for her friends).

I had to explain it to Connie, who never complained about anything -- she knew the rules and abided by them. She was never trying to be commercial, she was promoting reading, just like Oprah and her book club. Hey, maybe you'll see me on Oprah or Connie's show someday -- stay tuned!

Want a Haircut?

What about those hair demonstrations? Actually that's how they get new clients. No they are not being commercial, since they can't list the phone number of the hair salon they work at. What they can tell you is the salon's number when you call them up on their private phone. But you say that's not fair? Yes and no. Remember you have to call them at a non-business number to get their business number.

Yes, I agree, it is stretching the rules a bit, but that's the way it worked where I was. We even had people calling the station for the hair stylist's salon, which we all knew and would give out. And of course the salon itself was thanked at the end of the show -- so if you picked up the phone and dialed "411" you could get the number for the salon.

Business

There's actually one lady, one of the most popular on cable, that when you called her phone number, she had a message telling you to call information ("411") to get her specific business number. When you called the business number, you got offers of every kind to spend money and join a club. Again, she was getting around it because you had to call another number to get the business information. She, along with some other producers had (and still do) made a business out of Public Access -- maybe not a lot of money, but enough to more than pay for the cost of producing the show.

One Show at a Time

But the majority of people who have shows aren't making any money. They're having fun and getting little perks here and there. So let's be happy to start with a free haircut, maybe some flowers, maybe a free meal and some music. Then you can start making all those "Big Oprah Bucks!"

Chapter Wrap

Get a local sponsor for your show, give them credit and get a service in return. Use that phone and meet the people -- call a friend or relative. Don't think about getting rejected, think about getting free stuff! Keep trying until you succeed!

GUESTS

If you're not using a guest -- and you're doing a "Just You" show, then you can skip this chapter, but come back to it later as it will give you insight into expanding your show -- or trying something different.

Where Do They Come From?

In the big network world "Guests" have publicists, who try to book their clients on the most watched shows -- Oprah, Roseanne, Rosie, Leno, Letterman -- so they can be seen by the most people at one time. Basically they appear when they need to sell the audience on something -- a movie, a TV show, a book, an album, etc., and the host of the show gets more viewers; which in turn, boosts their ratings. When sponsors see higher ratings, they are asked to spend more advertising dollars.

Tony Robbins

Yep, the man behind the mega-selling *Personal Power Two, the Driving Force*. When he was out promoting one of his first books (I believe it was *Unleash the Power Within*), he was a guest on *Connie Martinson Talks Books*, a Public Access show. You might ask why someone as popular as he is (nowadays you can't turn a channel without literally seeing him somewhere) would be on Public Access. What have I been telling you since the beginning of this book -- please repeat it with me, "Everyone has to start small!" And back then, Public Access must have been one of the first outlets he tried. And also, Connie has a great following!

#1 Best-Selling Author for A Year

The author of the *Celestine Prophecies*, James Redfield, a national best-selling author, was on the *Looseleaf Report* in 1997 -- a show I worked on. Why Public Access? He could easily go on Oprah and be watched by millions of people. Well, why not Public Access? It's TV, it's local and most of all, it's FUN! Victoria's show was also well known and she had a great reputation.

New York and L.A.

When you're in New York or Los Angeles, any reputable TV show a person can promote themselves on, the better. Really! If you've got a popular show, in either of these two cities, then you are potentially being seen by the right people.

112

Why? Both are the major cities for entertainment in the world. Think about it, if you're going to be involved in TV, then you must get to one of these. Not to say you can't be popular in Chicago, Atlanta, Philadelphia, San Francisco, Dallas, Boston or even Mansfield, Ohio -- but eventually your show has to make it to New York or L.A.

Remember

Some previous stars have come through the studio where I worked: Dennis Cole (TV star of a variety of 1960s and 1970s shows), David Soul (*Starsky & Hutch*), Larry Wilcox (*CHiPs* -- have you seen the new episodes?), Monty Hall (*Let's Make A Deal*).

You might say, "big deal," they are all stars of the past. But they are still better known than someone who hasn't done anything at all. I heard a saying once, "Better to be a has-been, than a never-was." Or even, "Better to have used your talent, then not."

And these former stars are nice people who have something to say. Dennis Cole was talking about street violence and how he had lost his only son to a tragic and senseless crime. The story was featured on *America's Most Wanted*.

How About

Also, current actresses like Sally Kirkland, have been in the studio a few times. She was a guest on Arlene Peck's show, and also taped a show with the very popular Skip E. Lowe. Sally commented to me how "Skip really cared for his guests." It was true.

113

He had a lot of former stars on the show and would treat them with respect. What made him unique?

Skip's Style

He had a very recognizable style: an amber color tint to the faces, with extreme close ups and a black background. You have to watch him to get a feel for his show. He was very popular in Los Angeles -- you'd read about his show in local newspapers all the time. Actually he was something of a local celebrity. I directed one of his shows in the studio, with the actor Frank Pesce. Afterwards, Frank and I talked about possibly working together in the future. Again, Skip didn't start with "big stars," he made a name for himself by interviewing the former stars of old Hollywood, and in turn got so popular, stars from all generations were asking to be on his show.

More Hollywood

Hollywood Beat, a show I directed for two years, had guests such as: actor Chad McQueen (son of legendary actor Steve McQueen); film director and former president of the Motion Picture Academy, Arthur Hiller; film director Mel Stuart (*Willy Wonka and the Chocolate Factory* -- one of my favorites, remember the Oompah Loompahs?); and a segment I shot on location, with film director Peter Bogdanovich (*The Last Picture Show*). Not bad names as guests.

Soap Stars

Then there was a gentlemen who dedicated his show to Soap Stars, it was called *S.O. A. P. -- Soap Operas' A-1 People*, and was very popular with the local soap actors and shows -- *One Life to Live*, *Young and the Restless* and *General Hospital*. When he interviewed them, he was doing it as a host and as a fan -- you could tell he knew who they were and what they did.

You don't just call up *All My Children* and ask for "Erica," or *As The World Turns* and ask for "Dr. John Dixon." Remember, you should start with the lesser known names. Look for actors who have small parts on TV shows -- you can call the Screen Actors Guild (SAG) at (323) 954-1600 to get their agent's names and go through them to get the actor on your show.

Remember Phil?

Just last night I saw Phil Donahue on a local cable show called GNN (Gay News Network) out of Fairfax, Virginia. He was speaking at the National Press Club and was interviewed by the host afterwards. A guests such as Phil was great to see, since he was one of the pioneers of the daytime talk show style that they use today for Oprah, Leeza and Ricki, where the host walks out into the audience, instead of sitting behind a desk.

Patience

All of these shows have gained good reputations over the years, so don't rush in yet. Build your reputation slowly. You can get those guests that other shows do, but take your time. A year of building up your show's popularity is well worth the wait. Set your goals now.

Since you're just starting out you need to find a guest that is easy to talk to and is comfortable with you. Back in the "Talk show" chapter we decided on using a friend who likes music like you do. That's fine.

Billboard Magazine

Here's another idea. I worked at "Bpi" -- the conglomerate who owned *Billboard* magazine (you know, the one that has all those charts, the one people are always quoting about "being number 1 with a bullet!"). I started in the mailroom, where I managed the mail and myself -- but eventually started writing for the magazine -- then went into TV, and now I am coming back to writing again. I told people around there about my music show. Want to guess where my first guest came from? Yep. I asked *Billboard*'s West Coast Bureau Chief ("Editorial"), Dave DiMartino (who brought his friend Bill Holdship, the editor from *BAM* magazine along) to be my first guest on my premiere show. I got two for asking one.

Okay, I know I said not to book more than one guest, but Dave felt more comfortable with Bill along. So I had two knowledgeable guys to talk with about the music business. I worked with Dave and knew him a little, but still was nervous.

Silly Boy

I decided to do an opening where I was playing this big inflatable guitar to some music and dancing around, then all of a sudden realize I was on TV -- you know like when you got caught by your mom singing in the mirror with your hairbrush, or playing "air guitar" -- Mom didn't understand! Dave and Bill were laughing, and I did it at the end too. I knew it looked silly, but it loosened them up and I had FUN!

Listen

The show went well, except I kept cutting them off with more questions -- CLUE: listen to your guests, it's the most important thing to do -- remember, I mentioned it back in Chapter 7 when you went to the Producer's meetings? I didn't realize I was doing it until afterwards when I looked at it -- I didn't have good follow through and was constantly waiting for a pause, so as not to get caught in silence -- you know the kind when you first meet someone and the conversation stops and you start feeling awkward? Actors would say I wasn't living in the moment.

Some guy called me up and told me I was a "Jamoke" for not letting my guests talk enough. He actually liked my guests -- "you got great guests and you're not letting them talk." He thought they were very knowledgeable (hint, hint -- get guests who know what they are talking about), but criticized me for interrupting them. He was right.

By the way, I did learn to listen. It did take me a few more shows to get the hang of the "listening" thing, but when I did, it made all the difference. It was like I was having a conversation with this person, who I've never met before and we became instant friends. And you will get there too!

More Co-Workers

Yes, I did use many people from *Billboard*, from the editorial to the chart departments, as my guests. This showed the record company people that I was serious and would provide a show that would be advantageous and fun for their artists. Plus it allowed me the opportunity to get on the "Music Video Promotion" mailing lists -- they would automatically send me all their latest music videos to play.

Behind The Scenes

Also it helps to do something a little different and get guests that other shows don't get. For example, I invited behind-the-scenes people on the show -- music promotion people, retailers, writers, etc., who I thought had some insight about the music business. You usually don't get to see and hear these people on TV, and they are the movers and shakers behind the bands. And, like myself, people are curious about what it takes to make these bands "stars."

Radio

Did you know when you call a popular radio station, to request a song, if it's not on their "playlist" then they don't play it? Do you know how they get songs onto that playlists? You needed to watch my show and learn about radio promotion people -- they are a fascinating breed! And if you meet one, don't say the word "payola" -- uh-oh, now I've said it! Quick hide the book!

I heard one music guy from a very popular Top 40 station in Los Angeles say, "Music was something they played between commercials." To me that attitude sucks, but it is a reality. How about that as a guest? Could you imagine the audience's response?

If You Build It . . .

Word got around about my show, a few bands came on and the publicists and promotion people started calling. There's nothing like the feeling of a record person calling up and asking if you would "like tickets to a show, backstage passes, an interview with the band, their new CD and their latest music video!"

Rock N' Roll All Night

A highlight of my show was when I was cleared to cover the band KISS's inauguration in the "Hollywood Rock Walk of Fame" (outside the Guitar Center). Thousands of fans showed up and I was privileged to be allowed inside to ask them questions -- part of the press. It was fun asking them questions, with ten other people. The band had fun making jokes and talking about their long careers. I had been a young fan when they wore "the makeup" (yes I dressed up like them and sang all their songs like *Strutter, Christine Sixteen, Detroit Rock City* in front of the mirror, plus I even saw them in concert a few times). Man, they put on one heck of a show!

Hi Glenn, This Is Gene Simmons

No lie! Really! About a month later, Gene Simmons, KISS's bass player -- you know the guy with the extra long tongue who slept with over ten thousand women -- saw a copy of my show, called me up and asked me for a copy. So I sent him one. I still have his fax and phone number, but never use it, because I'd get too nervous and stutter

Pop Bop!

As my reputation grew, bands started coming to me. I interviewed them in the studio and on location. They would send me video taped i.d.'s for my show, for example Richard Marx *(Don't Mean Nothing, Children of the Night)* made an i.d. that said, "Hi to Glenn, Heidi and Baby Z."

You Know Art, Greg & Craig?

Everclear (*Santa Monica, Father of Mine*) was a band I interviewed in the studio, when they were just getting popular, and then on-location on the beach in Santa Monica. For the in studio interview we sat on the floor and did it in black and white (its a TCB thing). They were actually eating lunch while I talked to them -- it was very casual. They had fun and so did I. We talked about growing up, "eight-track players underneath the seat of a $50 Chevy Impala, High School reunions, bands like Kiss, Adam Ant," etc.

Later they went on to sell millions of albums and appear on *Saturday Night Live*. I wonder if Art, Craig and Greg remember me? If you meet them, ask them about the show where I did this fast motion stuff and we talked . . .

Start Small

Mine wasn't the only show that was popular, there were many other good Public Access music shows, and yours could be one of them, too. You just need to start small and work your way up -- it won't take long!

Remember, first find a local band or musician that you like -- it could even be a DJ on the radio or at a club. Invite them to be a guest, if they say no, then find another guest -- somebody will eventually say yes. You'll also be surprised to find there are some local bands that have a music video, or better yet, they might need one.

121

I'm a Friend of Glenn Darby

Use my name when you send information about
your show to the music video promotion people
listed in the back of the book. But don't be weird,
okay? Just be honest and cool.

Wrap Up

You've got a guest that is good, a subject and a
format, along with supplies, maybe a sponsor, some
music and studio time. Wow, you really are serious!

Let's move on.

PAPERWORK AND STUFF

In that studio package you got -- or maybe you
didn't -- there'll be more information than you or the
studio will need. For example, don't worry about
things like, Statement of Purpose, Statement of
Theme, Floor Plan or Story Boards. Reality is that
you only need to be concerned about two things: a
Rundown Sheet and the Credits. All the rest you
can verbally do.

Not to say you shouldn't think about a plan, just that
it's something you can say in a sentence or two to
certain people and they will do it. Here's what you'll
need to do.

Rundown Sheet

What is it? While they'll tell you it's required, I
mostly ignored them for the last two years as a
director. But you should get in the habit of having
one for every show. And if your show is basically

123

the same every time, you can use the same one over and over again.

You're in charge

Remember you are the producer, and the staff is there to help you. It doesn't matter if it's free or not, it's your time, efforts and energy. I hated when some of people I worked with used to say, "it's only Public Access, so what do you want." With that kind of attitude, you're sure to get a lousy show! Make sure if that director is bored with their job, at least you're excited about your show, and you bring that excitement into the studio with you!

Okay Kids, Get Your Pencil

Here's how to make your "Rundown Sheet." Take a letter-size lined piece of paper and write across the top, these five headings: "Total Run Time," "Segment Time," "Description," "Sound" and "Cues" -- making the Description column the widest. Now take a straight edge (a.k.a. a ruler) and make nice vertical lines to separate them -- all that elementary school training should be coming back. (I should note that some people like the Cues column to the extreme left, instead of the right -- that's fine.) And that's it!

Let's Fill It In

In the Total Run Time column, you write the start time of the show, which is "00:00." In the Segment Time column you write ":30" (thirty seconds) for the opening of the show. In the Description column you

124

write "opening of show, credits over a color background." In the Sound column you write, "CD -- track 1."

You have told the crew that the first part of your show will be used for your opening credits -- from your credit sheet -- superimposed over a color background of your choice, while a musical selection (track one) plays from a CD. So far so good.

Now come back to the first column, TRT, (for Total Run Time -- I'll be abbreviating some columns from here on) and underneath 00:00, you write ":30" -- since you have just completed the first thirty seconds of the show. Then in the ST column (Segment Time) you write ":30;" Description column, "host intros show and guest;" Sound column, "studio mics." Have you noticed, that no cues are needed yet? We'll get there. Read on . . .

Go back to TRT column and write "1:00;" ST column and write, "13:00;" Description column and write, "Host interviews guest;" Sound column and write, "studio mics."

Halfway Already

Now you'll need a halfway cue. The halfway cue is a visual reminder to you that you are half way through the show. Why? Because you only have twenty-eight minutes for your show and you need to know when to end it on time. If you decided not to have a break then, go to the TRT column and write "14:00," then slide across to the Cues column and write "Halfway."

PSA

If you want to take a break, for thirty or sixty seconds, have the director roll-in a PSA -- that's Public Service Announcement. You can ask the director to give you some choices before the show. PSA's come in a variety of lengths -- usually thirty seconds -- and themes.

Go to the TRT column and write "14:00" (halfway); ST column and write, "1:00;" Description column and write "PSA;" Sound column and write "S.O.T." (Sound on Tape), and in the Cues column write "Halfway." You'll probably start the PSA around 14:30 to 15:30, depending on how fast you can finish what you're talking about and tell the audience, "we're going to take a short break, and we'll be right back." Whatever you do, don't interrupt your guest, just wait for a natural pause.

Aaaaah

During the PSA, you can relax -- and ask your guest if there's anything they would like to talk about in the last half of the show. If need be you could actually stop the show and the tape from running -- have the director pick it up from the end of the PSA. Maybe you or your guest needs to use the bathroom, or one of you needs a cup of water, or maybe you need to check the parking meter. As much as I liked to run straight through a show -- live to tape -- we could always stop anywhere and I could edit it on the spot -- it was from all the experience I had from previous shows.

Second Half Kickoff

In the TRT column you write "15:00;" ST column, "12:00;" Description column, "Host continues interview with guest;" Sound column, "Studio Mics." Now you've got a full twelve minutes to finish your interview. But how do you know when you're almost done? Don't panic!

Here Come The Cues

You'll be able to get 5 minute, 3 minute, 2 minute, 1 minute and Wrap (:30) cues. So you go to the Cues column and write these in: "5:00," "3:00," "2:00," "1:00" and "Wrap", underneath each other. Also you might want to write in the TRT column the times at which you'll need them, "22:00," "24:00," "25:00," "26:00" and "27:00."

But wait Glenn, you said the show is 28:30 long. It is, but you'll need time to wrap up the show and run closing credits. Besides it will give you a little extra time to close the show -- don't worry if you're show is a little long or a little short.

The Big Close

Next to the TRT column, where you wrote "27:00," write ":30," in the ST column. In Description column, write "Hosts closes show" -- which gives you thirty seconds to do so. Sound is the same as above, "Studio Mics," and you already wrote in "wrap cue." Which is a whirly finger -- you might see it more than once if you're running long.

In the TRT column you write "27:30;" ST column, "1:00;" Description column, "run credits over set with lights dimmed"(I just thought I'd throw something different in the end for you -- I'll tell you more fancy stuff to do later for your closing); Sound column, "CD track 1."

The last thing you can write in the TRT column is, "28:30;" Description column, "Fade to Black."

RUNDOWN SHEET

Total Run Time	Segment Time	Description	Sound	Cues
00:00	00:30	Opening of show Credits over color background	CD Track 1	
00:30	00:30	Host intros show and guest	Studio Mics	
1:00	13:00	Host interviews guest	Studio Mics	
14.00	1:00	PSA	S.O.T.	Half Way
15:00	12:00	Host continues interview	Studio Mics	
22:00				5:00
24:00				3:00
25:00				2:00
26:00				1:00
27:00	:30	Host closes show		Wrap
27:30	1:00	Credits over set with ligthts dimmed	CD Track 1	
28:30		Fade to Black		

Credits

As far as your credit sheet goes, here's what you do. Take another piece of paper and at the middle top write: "OPENING CREDITS." Then start listing them. The show's title, The host's name (you), your guest's name and the topic. You can have yourself listed like "Hosted By Jane Doe" Or "Host, Jane Doe." Guest can be listed as "Today's Guest, Glenn Darby" or "Jane's guest is Glenn Darby" And you can put their title: "Author of the best-selling book 'How to Star in Your Own TV Show for $50 or Less.'" Find your own style and wording -- be creative.

Lower Thirds

Speaking of titles, you and your guest will need to be identified during the show -- you can even identify the show, too. It is called your "lower third" -- because it is placed in the lower third part of the TV screen. A simple one could be your name, then a thin line underneath it and your title, "Host." The same is true for your guest. But be careful with your guest's title, some studios don't want the name of a company, a product, a book, or anything you can purchase, used repeatedly. "Inventor," "Musician," "Author," "Actress," "Artist," all work just fine.

Multiple Titles

A fun thing to do for a lower third on *UCLA Bruin Talk* was to keep changing the title of the guest. Every time you saw the guest's name, it had a different title: "All time highest scoring player;" "Starting Player;" "Sophomore;" "Straight A

Student;" "Voted Most Valuable . . ." We had a blast with that one!

And The Show?

Like I said, you can even have a show title lower third. Why? Well, what if someone happens to tune in late and wants to know the show's name. Or want's to know the topic. A lower third can take care of that.

Put 'Em Here

Make sure you have the lower thirds labeled as such. You write them off to the side of your Credit Sheet -- usually they're down in the lower right corner, under a title "LOWER THIRDS." This way the technical person knows what they are. Usually they'll identify the people three times during the show. Some producers I've worked with like it more -- you decide and tell the director.

Warning Warning Will Robinson

CAUTION -- don't give out your home address or phone number -- people can make crank calls to your home. And you don't want your home address listed on a TV show. As much faith as I have in the human race, I would not broadcast where I lived -- it could hurt you. Just get one of those mail boxes -- not a P.O. Box, even though they are cheaper -- delivery services such as, UPS and FED EX, don't deliver to P.O. boxes. You used to be able to list it as a "Suite," but the U.S. Postal Service now

130

requires a "P.M.B." number for your private mail box. The people will sign for your packages and you can pick up your mail anytime. My wife and I pay about $15.00 a month for ours.

Check out the sample credit sheet:

CREDITS

1) Unzipped

2) Katie King

3) Unzipped guest is:
   Glenn Darby
   Best Selling Author of
   How to Star In Your Own TV Show
   For $50 or Less

Subject: Your Own TV Show (OPTIONAL)

CLOSING CREDITS

Then you write "CLOSING CREDITS." Which would be:
   Producer: _____ (you)

   Sponsor(s): _____ (remember from Chapter 10?)

   Music by_____, (remember Chapter 9?)

   Make-up by _____
   Hair by_____ (a sponsor, if possible)

For More Information (your address and phone)

LOWER THIRDS:

1) Producer: Katie King

   Host: Katie King

2) Sponsored by Wolfe on 2nd

3) Makeup by Marcy

4) Hair by Gib

5) Limousine Courtesy of Andre's Limo Supreme

6) Music by Sal Fortunata and the Stingrays

7) For more information or how to get an
   autographed copy of the guest's book
   write:

>   Katie King
>   P.M.B. 158
>   626 Santa Monica Blvd.
>   Santa Monica, CA 90401

or call: 310 280-3376

He Knows You're Home

And if you can't afford voice mail -- it's less than $10
a month --make sure you have an answering
machine on in your home. Don't answer the phone
after your show, let people leave messages -- they
prefer doing that anyway. You just don't know who
will be calling in.

That's all The Paperwork

Whew! There you have it, a completed "Rundown" and "Credit" sheet. You can confidently go into your meeting and then the studio.

If you're doing a simple talk show, verbally you can say the following: "It's a two person talk show, with a table and some plants; use a two color gel background -- blue on the sides and purple in the middle (they blend well together). Run my opening credits over black. I need a PSA at half-time and give me a halfway cue, five, three, two, one and wrap cue. Run my closing credits over the set with the lights turned down. Here is my rundown sheet, credits and lower thirds, please identify us at least three times during the show. Also here is my 3/4" tape for me to take home, please use a stock tape for you to air and here is my blank VHS tape. Here is my CD for opening and closing music, it is track one." That is your whole show!

Do you see how easy it is when you know what to say? If your show is a little more complicated, you can still use the basis what I covered here and word it so it fits your show exactly the way you want it.

Checklist

Subject; Format (Talk Show, Demonstration, etc); Music (CD or cassette with the rights cleared); Sponsor (free haircut); Guest (a friendly one); Studio time booked; a 3/4" videotape and a VHS tape; a "Rundown" and "Credit" sheet.

Budget Check

Start with $35.00 for studio time, $6.00 for Eco
Tape or $12.00 for a new one and $1.00/$2.00
 for a VHS tape, equals $42.00 to $48.00. We're
looking good! The $42 budget will allow you some
food and parking money -- plan ahead.

Okay, let's talk about make-up -- yes guys, you too!

MAKE UP AND WARDROBE

No I'm no superstitious! There is a Chapter 13.
The town where I am now, Santa Monica, doesn't
have a 13th Street -- it's called "Euclid." Plus I'm
writing this chapter on the 13th day of the month.

Somewhere in that package they gave you at the
studio is a section on clothing and make-up. The
lights in the studio are 1,000 watts (1K) (remember
those $35 light bulbs I mentioned?) and they reveal
everything. Video is not kind to the skin. It is not
flattering. Film, like in a movie, can be very nice to
you. Have you ever seen a movie, then see one of
those interviews or behind the scenes shows,
where the actor looks completely different? Besides
the hair and make-up artists, it's different because
it's shot on video.

Okay Guys

Men, all you should wear is a simple base, and if
you sweat, a little powder will do nicely. There are a

plethora of brands that make bases. You might want to use a hypo-allergenic one that won't irritate your skin. If you're light skinned, don't get a dark base -- it will look too funny. Just get a shade that matches your skin tone. Ask a female friend who you think looks great, and doesn't use a lot of makeup, what she thinks. You don't want people noticing your makeup.

One producer, a doctor, came in wearing an orange base and it looked like he was wearing orange make-up on the show. No one ever told him how silly he looked. Another guy came in wearing too much make-up (eye liner, blush, mascara, etc.) and he looked like he was wearing make-up. The right combination will complement your skin and its features, not take away from it.

Real Men Don't

Okay, think of the toughest action stars in the movies: Jackie Chan, Bruce Willis, Sylvester Stallone, Wesley Snipes, Arnold Schwarzenegger or Clint Eastwood. They not only wear makeup on the movie set, but also when they do talk shows.

"No they don't, I can tell," you say? Ah-ha! That's the sign of a good make-up person. Everyone on broadcast TV wears some kind of makeup. It doesn't matter if its news, sports, talk shows, sitcoms, infomercials they all wear makeup.

Alright I Won't Force You

Often men won't bother with makeup for Public Access, because they don't think its important. But wait until you see yourself on the tape with those

1,000 watt lights beaming on you -- they will highlight every blemish and imperfection (not to say you have any). Bottom line is, that very few people can get away with not wearing makeup (the always tan kind) no matter what your skin tone is.

If you want to skip it the first time, that's just fine. It'll give you something to work on for the next few shows.

And Women?

Unfortunately women have been conditioned, by men, all their lives, to believe they look better in make-up. Since you wear it anyway, put some extra on for the cameras -- a little bit more won't even show. My wife, who doesn't wear much makeup -- I think she's beautiful without it -- puts it on heavy for the cameras and it still doesn't look like she's wearing much at all. Base, blush, eyeliner, lipstick and that's it. Stay away from the "Tammy Faye" look -- unless it's your style. A good rule for women is that whatever you wear everyday is fine, but add a little more to compensate for those 1,000 watt lights. You'll discover how you look when you look at the tape. But don't be too critical!

I Don't Have A Thing To Wear?

Clothing? You might have heard someone say not to wear white, red or black, but you don't know why? Let's start with white. The camera sees darkness and brightness, and the iris (like your eye) must open and close according to what is being reflected into it (light bounces off objects). If you're wearing a white shirt (very reflective), then the iris must be closed to cut down on the light coming in.

The problem then is that your face will appear dark. And if you have a dark complexion, you won't be able to see your face. Most Public Access facilities don't have a special lighting person, like network shows, who can light your shirt separately from your face. If you do wear white, don't wear a lot of it and cover it up with another color -- a white shirt can be covered by a blue or gray jacket.

Black is Black

Black isn't as bad as white. Black doesn't reflect much light (no detail), so the iris has to be opened up (or you need more light), which will wash out other colors, including your skin. You can compensate for black by asking the director to bring up the "black" (it is a button on the camera) level a bit. I've had people wear black suits, sit in front of a black curtain -- talk about a nightmare! They looked like floating heads.

A Little Red is Alright

Red? It bleeds. Really. When you wear it, it can spill over, outside it's border. A red jacket might look softer around the edges as it extends to other areas. Also when you're dubbing tapes, or editing, it is usually the first color that degrades.

Stripes?

Stripes are a little deadly, small ones will create a rainbow like effect -- that you can't get rid of. Some men's shirts and ties are made like this and you'll see a little rainbow like effect, which is very distracting.

Also watch the horizontal stripes on dresses as they tend to show all the bumps and curves. The vertical ones are usually okay. Yes it is true about the camera putting on ten pounds -- it's funny but I've seen it all the time. You just have to experiment with your clothing to see what makes you look the best.

Keep It Simple

So what do you wear? Pastels, earth tones -- blue, grays, greens, purples. Combinations are good. A little black, red or white is fine. Depending on your theme, clothes can standout (fashion show), distract (too much going on) or complement your show. Remember in today's multi-channel TV sets, you only have a few seconds to catch their attention, so wear it well.

And if your guests asks about what to wear, just tell them whatever they are comfortable in, but also use my guidelines to advise them.

Done Already?

That was a fast chapter, wasn't it? Basically you will figure out what and how you look best on TV.

Ask a friend, who's opinion you can trust, to watch the show and tell you what they think of your appearance. You can learn a lot from watching other shows too.

Now that you're looking good, let's talk about the Big Day!

THE BIG DAY

Well it's finally here, the moment we've been waiting for. You have everything you need and your guest has confirmed the date and time. Plus they know where the studio is -- you gave them directions right? They know to be there before you start taping -- right? And you told them to allow at least an hour's time for the taping -- right? You are good!

Pick Up The Phone

Let's say you're taping from 3:30 p.m. to 5:30 p.m. this afternoon. The studio has a taping day of 8:30 a.m. to 10:30 a.m., 10:30 a.m. to 12:30 p.m., 1:30 p.m. to 3:30 p.m., 3:30 p.m. to 5:30 p.m. and 6:00 p.m. to 8:00 p.m. You'll notice there are shows before you and after you.

Before you rush over there, call them up at about 2:30 and make sure the show before you is running on time and not late. They should know this, which gives you an hour to get there.

TIP: Studios usually like you and your guest to be 15 minutes early, but they don't really start taping until at least a half hour into your time. You can be a little early, but your guest doesn't have to wait around for half an hour.

You're There

You called them up and they are running on time. You get there at about 3:15. Ask someone who works there to let the director know you're waiting in the green room. Someone can call the booth for you. Or they might tell you to pop your head in the studio and let the director know you're there -- this might be your director, or you can have someone find your director. I remember days when I directed all the shows for the entire day. If they're somewhere else, have someone find them and let them know you're there.

They might be busy, depending on the show, so don't feel weird if they don't pay attention to you, just as long as your director knows you're there waiting in the green room or makeup room. They'll probably have a monitor on in the green room, where you can watch what they're taping, so you'll see when they're finished.

You're Early -- They're Late

If you call the studio and they're running late, get there on time anyway, because just you being there is pressure enough for them to hurry up; otherwise, they'll think they can tape all day. But let your guest

142

know, and call them when you get to the studio. Most shows ran over only a maximum of thirty minutes -- but usually ten or fifteen.

Your Guest Arrives

The guest arrives and you explain to them that they're clearing out the show before you. No problem, since they might need to use make-up, the bathroom or get a quick snack.

Also you'll want to talk to your guest about some questions you'll ask, and if there's anything special they want to talk about on the show -- holding up a graph, book or article. Let the director know this ahead of time so they can tell the camera person. You might have done all this when you met with your director earlier in the week, but go over it again anyway because you don't know if they remembered.

Release Form

Have your guest sign a release form now. Don't wait until after the show -- they might change their mind or have to leave. Explain to them that's it's a standard form, and you're not going to exploit them in anyway or make money off them (and if you do you'll negotiate later) -- it's just Public Access and we're having fun.

You're Up

Your time is here. They're clearing out the show before you and an intern or the director comes in. Here's what they need to have and know based on a simple talk show format. You tell them it is a talk show, one host, one guest -- two chairs and a table, and a few plants would be nice. Remember the rundown in the previous chapter?

Your Background

And here's the key to the look of your show: tell them you would like a background color that complements what you are wearing (try not to get a curtain -- it's too cheesy). Get a two-color background -- one color on the sides, behind you and the guest, and one color in the middle background -- so that camera two (the middle one) sees one color on the wide shot, and cameras one and three see the other.

Some directors tried to pawn off the same color for each show. One day I lit the 8:30 a.m. show and every show after that used the same exact lighting. Don't let them do it! And don't get talked into a blue background because it's commonly there, especially when you're wearing blue -- it looks awful. Try more complementary colors. For example, if you're wearing a blue, use a light rose color. Blue as a background works well with everything, except when you're wearing blue. Also try to break up the solid color with some other color gels. A blue background can be broken up with some subtle yellow and red streaks (close the barn doors). A rose background can be broken up with yellow and green streaks.

Be Creative

Tell them your background color of choice, then ask them to add some subtle colors to make it unique using their creative sense. A good director will work with you, a bad one -- well, that's when you take over. Don't let someone tell you what to do. This is your time and your show, they are there to help you. Remember, they all have someone to answer to. Don't be afraid to tell them you don't like something, nicely.

Give It to Them

The Rundown and Credit Sheet will tell them just about everything. Also give them the 3/4" tape and VHS tape, and ask them to tape the show on one of their stock tapes. Give them your CD and tell them which track is for opening and closing music -- this is on the rundown sheet, but you should emphasize it.

If you're using a PSA, ask them to have a few ready to choose from -- tell them your values -- the environment, child abuse, drunk driving, etc .
The reason why you should tell them what you want in addition to what is written down, is because it reinforces the plan of your show and they will understand it better.

The director or intern leaves, while you and your guest talk a little and get comfortable.

Make-up

Uh-oh. If you called ahead and found the student make-up person to do your make-up, for credit, you could have it done before the guest arrived. Otherwise, now is the time to get it done, then have your guest get made up -- if they didn't already have it on. While they are in there you can be checking out the studio. Note: some guests won't want make-up. That's okay too. Suggest it, but don't force the issue.

Usually the studio has a list of make-up artists available -- if not look in the phone book -- or one of the TV industry books like the "L.A. 411," under "make-up" schools (preferably theater or TV based) and inquire about their students. Although it sounds risky it can work very well for both parties, but try to get a recommendation first.

Pop In

It will take them about half an hour to prepare the studio for your show. Don't wait, after the director takes your tapes, check in about fifteen minutes later to see how they are doing. Introduce yourself to the crew. They might have questions for you about what style (fonts) you want for your credits. If you like what they have thank them, if not, pleasantly ask if they have anything else -- remember the more time they take to show you something, the less time you have for your show -- choose quickly!

Your Team

Ideally, you'll have a director, a TD, a credit person, an audio person and two or three camera people. They all answer to the director -- who is sometimes called a "Production Supervisor." I was a "Director/Production Supervisor/Instructor/Intern Supervisor." That title and $3.95 would get me a Protein Berry Pizazz from Jamba Juice.

Have the audio person check your CD for the correct song. Make sure they know to use it for the beginning and the end (or where you specify).

Ask the director about the PSA -- they should have a couple ready for you to chose from -- studios have dozens of PSA's to look at. When you told them your values, they should have a few that will fit your show.

Check the Lighting

Walk out of the control room and into the studio area to make sure they are setting up the right color cyc (background). Ask if you can sit where you'll be sitting to see how you look. They can show you how you look on the studio monitor. Check to make sure your light is not too harsh -- they should be using diffusion (some sort of fogged gel or spun) on your key light -- the one that shines in your face. If not, ask for it.

Oh, and about that "studio monitor." Unless it is absolutely necessary that you have it on during the show (like you need to see roll-ins), please turn it off. It is very distracting, especially for the guests.

Your "back" light -- the one that shines on your shoulders and head -- should not be lighting up the top of your head and not so bright that it is very noticeable. TIP: Watch the evening news and you'll see a medium light on the person's shoulders. They should also use diffusion on it. (Lighting note: all white light should have some kind of diffusion -- spun or a fogged gel on it -- don't let them bring the dimmer down because it can cause you to look orange.)

Your "fill" light should be just bright enough to wash out the shadow from the "key" light, yet not bright enough to cause another shadow. You'll notice that News anchors have two shadows underneath their chins -- look closely on their necks. Try to avoid this for your talk show. TIP: Watch one-on-one interview shows like *Dateline, 20/20, 60 Minutes*, to see how they are lit.

Basically, you'll end up with what is commonly referred to as "flat lighting" -- it is TV talk and eliminates all shadows. But maybe you like a little shadow -- it can be your style. I would try to encourage people to get away from that flat look and experiment with shadows. Now it wouldn't work for an Oprah or a Rosie show, but when you are being intimate, it could work very well to use less lights. Remember, experiment and have fun!

Leave 'Em Alone

If you're happy with the way it is going, leave them alone -- they hate producers watching them work -- and return to your guest. If they don't come get you

148

in fifteen minutes, pop back in and check on them. Always offer a friendly word.

Last Minute

Okay, your time is here. The director gives you the big thumbs up. As they are seating your guest, you'll want to check your credits for spelling errors. Do this in the control room if possible. If not, have them use the monitor for you to check them -- either way, but DO check them! They are your responsibility. Most of the time I could go back and fix an error on any show, but it could turn into a long drawn out process. Take the time before and check those credits -- please!

Double check with the director about how you want the show to go. If you want a PSA ran at half-time you should've picked one you liked.

Cues

Your guest is seated with a microphone on. Your credits look good. A camera person or the director shows you the hand cues. "Half time" is a closed fist, five fingers for a five minute cue, three for three, etc., and the whirly finger for the wrap up. They'll be given by your camera person -- either camera one, two (rarely) or three, depending where you are seated. Stage left looks at camera one, while stage right looks at camera three -- camera two is a set shot and usually doesn't have an operator.

Lookin' Good

Ask to check your camera shot in the studio monitor, along with the other shots too. Ask the director to vary the shots. Over the shoulders are good, and if the camera person is experienced, a slow zoom (push-in) on air is good, and the reverse (a slow pull-out) is also cool. Keep it simple, but creative. All TV directors like to think that they are being creative, and when you tell them that they feel good. But you have to be firm about letting them do anything they want. They can ruin a show by trying to be too artsy. Camera two is a boring shot, so let them know you're not fond of the set shot -- use it very sparingly.

I got to the point where I would ask a producer what shots they'd like, some didn't know, so I would run them by him or her and quickly show them what looked good. They took that with them and started telling the directors at other studios what they wanted -- good for them! Don't be shy about telling a director what you want, because you are the producer and you hold the ultimate responsibility of how your show is going to look -- not to much pressure? Relax and have fun.

No One's Perfect

No matter how good a director is, there will more than likely be a move or shot that you don't care for -- that's okay, don't be upset, people do make mistakes. I've been responsible for some bad shots as a camera person and as a director. You might have an out of focus shot or a bad zoom. You can see those little flubs on the "News" all the time -- and they're broadcasting to millions live. What do

150

they do? They keep going, and that's what you do. But make a note of who it was and if it keeps happening, try not to have that person (director or camera) do your show again. But don't make a big deal out of it -- I think I'm actually writing too much about it myself!

I Can't See

The lights are brighter than you thought they would be. You're nervous -- you can bet Leno and Letterman still get nervous before they go on. There's some talking, a few people running around, a lighting adjustment. Again, you should've given the okay on your shots after you've seen them. You can tell if you're too bright or dark -- don't be shy -- say something, politely, "Can you make me less bright or a little brighter."

Quiet Before The Storm

Someone yells, "quiet" and then, "Stand by!" Well this is it. The silence is scary. Then your floor director and/or camera person holds up a hand and counts down, "five, four, three . . ." (two and one will probably be silent). They will hold up two fingers, then one -- then they point to you, a red light goes on above your camera and you say . . .

"Hi and welcome to my show." Whatever you do, don't stop. If you stumble, catch yourself and keep going. Don't apologize to viewers, because they'll turn you off. Rather, make fun of a mistake -- forgot your guests name -- "OOPS, there goes that memory thing again, must be all these electrical lights effecting my brain . . ." Have fun! If you really mess up at the beginning you can ask to start again. But once you're into the show, just keep it going.

Whew, Time Flies

You'll notice how quiet it can be, and how fast the show goes. Before you know it, it's half time and you take a break. The PSA looks good. You pick up the second half and then you start getting cues. A good director will give you cues when you're not on camera -- or should I say, your camera is not on you. When you get the one minute cue, don't ask anymore questions, simply follow up what the guest has said. And if they're going on too long -- the camera person is going crazy with the whirly finger -- you can always use the line, "sorry, but we're running out of time -- maybe you can come back on another show."

It's Over, Almost

Thank the guest. Close the show with something simple, like, "thank you for watching, see you next time on_____." Now they run credits. You can leave -- unless you're doing a show where they run credits over the set. Oh yeah, that cool effect? Okay here's what you do for future shows.

Something Really Cool

For your closing -- have the director put your white lights (key, back and fill) on one dimmer group/bank and your cyc lights (the background) on another. When you close the show have them bring down the white lights and you'll appear as a silhouette. In your rundown sheet you put it in the description column -- run credits over set (camera 2) in silhouette. If they say they can just iris down -- don't' let them, it is cheesy. You can also do this silhouette effect for the opening, but they must time it correctly, as the lights come up, you start talking. Try it, it looks good.

Another cool effect is to have them bring the white lights down 50%, at the end, and run credits, while they dissolve from camera to camera, as you and your guest are silently talking, in a semi-silhouette.

You can do a version of this silent-talking under credits, for the opening, but it is very tricky to do. Some people will just keep talking at the beginning for about five more seconds after their cue, then introduce the show. This sometimes would freak out a director. I'd be talking to my guest and the camera person would be cuing me over and over again, and I could hear the director saying, "tell him he's on," and once the camera person said, "You're on," I just looked at the camera and said, "Ooh I guess we are . . . hello and welcome. . . ." Have fun!

All Done

The camera person yells, "Clear!" Now you can leave -- OOPS, don't get up without having your microphone removed first. I can't tell you how many times a person got up too quickly and ripped the mic off. After awhile the mic broke -- they cost about

$150 and they could charge you for it. That would really cut into our budget, so please take a moment to make sure everything is off you before you get up. I know you'll be so excited from the show, but you'll need to breathe. Good!

So, How Was I?

Duh! PLEASE don't ask anyone how you did. They are never going to tell you that you were terrible, even if you were! And you don't need their approval anyway. They'll probably give you the same cliché, "you were great."

Instead, let someone approach you with a good word. If they liked your show someone will tell you. They might ask for more information, or maybe have something to share with you. Be open and polite. Some interns even talk to the guest -- that's okay too, let them while you get your tapes. And don't forget to thank the crew -- that is a great gesture. I've even seen some guests introduce themselves to the crew before the show.

Really, How Was I?

When a producer came into the control room, they would ask, "so how was it?" When I first started working I'd tell them, "it was great." Then later, "I'm sorry, I was so busy making sure the show ran well, I didn't get a chance to listen." It seemed to be more of a self-esteem and ego issue with most.

154

Eventually I got to a point where if they asked, I'd pull them aside and point out what was good, and what they could improve on. It would always be from a technical view: "try to have a roll-in with you, it could add to the show." Producers do like to hear how they can improve, but only after they get to know you. So after a director gets to know your show, start asking them technical suggestions -- in private, NOT in front of your guest -- about how to improve it. They can have some good tips. But remember you are the ultimate judge of your own TV show and you decide what looks good and what doesn't. Don't be afraid to experiment, but don't be afraid to accept advice too.

In the Lab

One producer tried a show using a freestyle camera look that could have worked, but didn't for his show. It was creative and daring, but not for the kind of show he was doing. You could see the camera people walking around the set and doing those moving angle shots. It took away from the show because it didn't fit the topic. Still, it was a good experiment.

What's That Roll-in Thingamajiggy?

Oh that? That's what you use when you want to show some footage. If you're talking about a band, you would bring a "clip" of them performing. Sports? Video tape highlights from a game. Medical show? Footage from an operation (ugh,

155

please, no blood). It is something that shows people visually what you're talking about, that you can't do in the studio. It helps a show because it takes people out of the studio and into another place.

Back To Your Show

You're finished with the show and you are in the control room. Now, what about your tapes? You should have a VHS copy of the show, for home viewing; a 3/4" master copy for you to take home, and make copies if need be (for bicycling together studios and a VHS copy for yourself). You can always recycle it later when you need a tape. If they made a 3/4" stock tape air master -- for them to use for broadcast, and then recycle later -- take that one to the "Traffic" person, which is in the next chapter.

Be Afraid, Be Very Afraid
Maybe a Little?

Wait! Before we move on I need to tell you a little horror story (no blood, I promise!). When you get to the point of doing two shows at a time (in the two hours or so allotted), and you take a break between the tapings, make sure you physically take the master 3/4" tape from the first show and put it away. Here's what happened to me once.

I was taping two interview/music video shows at United Artists in Van Nuys, CA. The first show guest was Starr Parodi -- she played keyboards in the "Arsenio Hall" show band. She had her own album out and her own music video. I found her through a friend of a friend who knew my show.

156

She was a great guest and we had fun. I'm still in touch with her and her husband Jeff, who together do scores for movies.

We finished the show, I thanked her and she left. I checked in with the supervisor and he told me everything went well. We set up for the next show -- the guest didn't show up, so I just did a music video "Just You" show. This is a good note to remember, if you're doing multiple shows, and the guest doesn't show up, you might want to be prepared to do one by yourself -- if your theme will work for it.

I played some videos and I ended the show with a trivia question. The prize was a CD (I started giving them away to the crew). Then I went to get my tapes. The supervisor comes to me with a "someone-just-died" look on his face. I asked if everything was okay. He told me, "we taped over your first show (3/4" master tape) with the second show." I said, "What?"

He said, "we forgot to take the tape out of the deck and accidentally taped over your first show (the one with Starr Parodi), which now has your second show on it."

DUH!

I didn't have a backup 3/4" master tape running --
so what could I do? Call Starr up and tell her all that
spontaneity was for nothing? Ask her to come
back, with her busy schedule? No!

I luckily recorded the show on a separate VHS tape,
which they didn't tape over. Although the VHS
wasn't as good quality as the 3/4" tape, I was able
to leave there with it and bounce the VHS up to 3/4"
at another facility. Whew!

This story just goes to show you how careful you
need to be when taping shows. I told the story to
the people at Century and they couldn't believe it!
But ultimately I was responsible for getting my
master tapes out of there. I laugh about it now, but
pass this own to every producer I talk to.

And yes, I went on to tape more shows at United
Artists, and had fun. We even laughed at the time
they taped over the show. Everyone makes
mistakes, but if you take precautions and learn from
others, like me, you'll be better off and make fewer.

So, on to airing your show. . . .

AIR TIME

The show is done, your guest is gone and you're exhausted. Now it is time to pay your $35.00 (if they charge), and book your show. You go to the office and meet with the "Trafficking" person or the director might follow you up, take your money and book your show. Let's say the director left and you're with someone new.

Everyone Can Do It

"Trafficking" is the term we used for scheduling shows. There would be a person assigned to this area, yet everyone there at the studio could do it. I was asked to do it during my first year of employment at Century Cable (it was an "office politics" thing, I won't bore you with it), while they looked for a replacement. In addition to my directing, teaching and training, I covered the trafficking department as well. Basically, it was the most loathed job in the department.

With a Little Help From My Friends

So how'd I do it? No, I didn't work ten or twelve hour days. Since I was in charge of the intern program, I had the interns help me, while training them how to do it. Which of course made it easier for me, and the interns liked it because they were learning something new. No one had ever let them schedule shows (they didn't trust them). Instead of me sitting at a desk listening to people ask, "what is the best time?" "How come I can't get the time I want?" "Why does he always get the best time?" I'd have my trusty interns do it most of the time, and they'd do it very well! Eventually some of those interns got offered jobs in the department.

A Funny Thing Happened

The interns and I started running a new "Music Video/PSA" show five days a week. You see, the traffic person was responsible for rerunning any show needed to fill in the empty time slots (more later). We had fun making up the shows -- the interns would take turns as hosts and picking music videos and PSA's.

Moral of the Story

Actually two. One, be nice to the person scheduling your tapes, they might have just had a run-in with an irate producer, or maybe they're having a bad day -- you know how that can be.

And two, if you're offered a "trafficking" job, think about accepting it -- it could lead to a better one.

160

We went through many people in that position, but some hung in there and directed shows while working in the department, which eventually turned into full time directing.

Tapes From Beyond

The "Traffic" person who schedules your show can be booking over fifty shows a day. But Glenn, you said they only tape about ten shows a day. Where do the other tapes come from? Remember you can tape a show at another studio, then bring it over for airing. Then you have the mail-ins (sometimes twenty a day) which means the person could be extremely busy when you come in. Either way, again, be nice and patient, allow some extra time, and you'll get the schedule you want. Okay? Let's schedule it!

Finally, When to Air

You can choose anytime you want. DON'T ask them what is best. PLEASE! I got to the point where I told people I didn't get involved in picking time, because if I said a time was good and no one called their show, then they would blame me. Unfortunately this would irritate people more. So then I started saying, "anytime was good, someone was always watching." That didn't work either.

What's Available

They are supposed to be available 24 hours a day (it is in their agreement with the city and can be used to your benefit), but Century, had limited it to

Monday through Friday, from 12 noon to 12
midnight, and Saturday and Sunday from 8:00 a.m.
to midnight -- which was more than adequate.
However, if you wanted to air it at 12:30 a.m. in the
morning, you had to put it in writing and submit it,
and then they would have to put it on. But they
wouldn't go any later -- you were told that that's all
you could have. Wrong!

You're a Trouble Maker

But be warned, when I worked there, if you wanted
to air your show at 2:00 a.m., you'd be causing
discomfort and they would label you. They had a
verbal list of "troublemakers" and would avoid doing
them any favors. For example, if you were calling to
find a cancellation, and you were known as a
"troublemaker," they wouldn't tell you about one if it
was sitting right in front of them. One week we had
about 50% of the shows cancel and no one
bothered filling them in -- unless a friend needed
one. Things might have changed -- most of those
people are gone now, but I was there recently and
they were still not filling in some prime studio
cancellations. Bottom line? Public Access is there
for you when you need it -- it is your right!

Opposite "ER"

If you read this book before booking your tape, then
you'll know to pick a time before you go in. Some
people had a theory about certain times: prime time
was supposedly the best because that's when most
people were watching *ER, Spin City, Friends,
Seinfeld, 60 Minutes*, etc. But if they were watching
those networks shows they weren't watching Public

162

Access! Producers thought during a commercial break the audience would stray over to them. Great if your show was on Channel 3 (between CBS and NBC), but what if it was on Channel 13, or 26, or 57 or 77? Don't count on people coming by from a popular show to yours.

Soap Time

And what about the daytime? Plenty of people watch the "Soaps" (everyone knows who "Erica Kane" is -- don't they?) and other talk shows too -- Oprah, Rosie, Leeza, etc. People do watch TV in the daytime -- that's why you have air time then, and it can be a great audience! But make sure your theme fits in with that audience.

Time & Theme

From my experience, sometimes picking a time that is relevant to your show will help. Sunday morning worked well for religious programs. Afternoons were good for kids, who would come home from school and watch TV. A night time show about clubs isn't really going to catch on at 3:00 in the afternoon, but at night you might have a better chance when more people are home from work and looking for something to do. After 11:30 p.m. the "Late Night" adults were supposedly up surfing for something different (Remember *Colin's Sleazy Friends*?).

The Truth

Reality is that no one knows a good air time for your show -- you just try it and find out. Look at all the prime time shows that come and go. They probably have a about thirty new ones per season and about five or so make it through the year.

If someone tells you the best air time, they're fooling themselves and you -- don't fall for that "prime time" slot. Use the guidelines I've given you: for the most part, time of day corresponds with a type of audience.

Hello Out There

Now that you picked time, how will people know when it's on? One guy, "Skip," went so far as to rent those bus benches with an advertisement (picture of him, stations and times when to watch) for his show. Did it work? People did notice the bench and I'm sure some people tuned in, at least out of curiosity. And when they did, if they liked it, they'd keep watching.

Some people, who didn't have much of a budget, would just put up flyers in local supermarkets. Any word you can get out about your show, the better. It's like anything, people don't know about it if you don't tell them. Of course, they could just stumble upon it by chance. And how will you know if more people are watching?

Call Me, Please!

Remember that phone number you put at the end of the show? If after you air your show and no one calls in, ask yourself why. Do you want them to call you and tell you how wonderful you were? Sure you do! Just as I want you to read this book and tell me how it has helped you get your show together. But that's exactly what a lot of producers want and don't get. That's the wrong reason to do a show.

Give 'Em A Reason

Why should a person call your show? Are you offering some valuable information to them? Are you offering a prize? A FREE CD? Make sure you have a reason for people to call in. Why would you call in? Think about it. People have to be motivated to pick up that phone and call you (in Chapter 19 I'll tell you about shows that get twenty or more calls). Give them a reason and they'll call.

One quick note: addresses for the most part were a waste of space -- people couldn't write the information down fast enough. But now with "The Web" you can put your email up and people can send you goofy messages!

I Want To Ride My Bioyole

"Bicycling." Actually you probably won't use one for this. What it means is that you take your 3/4" tape around to different stations for airing -- don't use your master. In L.A. you can tape at Century and air it on Media One, TCI, Cox, etc., or send it off to New York or Austin and they'll play it for free (check

165

the Reference Section for numbers, addresses, and a "hot tip.").

I used to have my show on in Austin, TX, in the Philadelphia suburbs, New York and all around Los Angeles. It took time and postage, but you can get your show on around the country by simply calling the stations and requesting information about their Public Access facility.

You may want a resident of the town to sponsor your show. So what you do is ask if they have a trade-off, where someone there will air your shows, in exchange for you airing their tape where you live. Here's what to write: "Producer in LA. looking for a producer in Austin to sponsor her shows in exchange for sponsoring your shows here." Or run an ad in the local paper. It takes some effort, but you can be on nationwide!

Every major city in the USA has a channel for you to air your show. There are even channels in Europe for airing, but you have to network with people around the station about leads for airing your show around the world. You could even join or start a local "Access Users Group."

Rules: Behave Yourself

Earlier I mentioned I'd talk about the rules and regulations for air time. Again I base these on my time at Century, where you were allowed only four shows at a time on the books, and those books would run for a year. When I say books, it was actually a computer based system, that would tell the person booking your show, if you had more than four shows.

166

You could book your four shows at anytime that was available, but only one show per week. When a show aired and you had three left, you could go in and give them another one, to try and keep it consistent. Unless you were on at an unpopular time (weekday afternoons), you'd probably have a break in your schedule.

Not all studios operate on the limit schedule -- some are glad to get shows and will run as many as you can give them for as long as you want. Some stations even repeat your show four or five times! Check with them when you "bicycle."

Exceptions

Dr. Susan Block was allowed six shows running at a time, because of the adult content of her show, which the station preferred to be on late at night (after 11:00 p.m.). It was a compromise that worked in her favor, because she almost never missed a week. You could see her show around 11:00 pm to midnight, the same night, every week. HBO eventually picked it up for a time. Again, the late night time worked for her and her audience.

Rerun

No, not from *What's Happening*. As mentioned briefly, "Reruns" were shows that aired once already, and were used to fill in unused time slots. For example, when the schedule, say for Tuesday, from noon to midnight, had unfilled bookings at 1:30, 3:00 and 4:00, the traffic person will fill those times with shows that had already aired once. What shows to use were quoted to be "random."

But that was almost never the case, from my 5 ½ years of working at Century. They would be filled with a friend's show; in fact, one lady had about twenty shows booked! It was her friend who was doing it as a favor. He was not the traffic person, but like I said, anyone could book the shows.

At one point they were double and triple booking shows. Which meant when you scheduled your show, they would schedule it once or twice more. This got a bit confusing since the tape was put on the already aired shelf and sometimes forgotten about. It was solved by putting two stickers on the side to remind the traffic person it had to go back into rotation. It also got out of hand because word got around that you could schedule your show as a rerun right away and people started demanding it. A lot of the multiple bookings stopped for awhile, then started again. The producers themselves, who were over booking were actually complaining about the practice. Some producers felt since they had been on the channel for many years they had the right to extra airings and no one else could. Yes this is true! Public Access had created "Air Time Monsters!"

I understand they've cracked down on the misuse of reruns and are now being less biased and more open to the "random" picking of reruns. Yours could be one. People are human and if you don't get a rerun, ask nicely and maybe they'll play it again. But don't be hurt if they don't.

Old Shows Never Die

Once you've aired a show and it was done, (including reruns), you weren't allowed to air it again, ever. But that didn't stop the producers from trying to air them again. You see, no one had time to screen all the tapes that came through that studio. For a time we would screen them, but that got to be time consuming. So producers would bring in old shows, with the sticker of the old date and air time. Of course we would say, "Sorry, but this show aired here already -- see this sticker, it's ours."

So they started putting new labels on the old shows and airing them. Did they ever get caught? You had a better chance at winning the lotto than getting caught airing an older show.

And if they did? They'd be told not to do it again. Or another producer would turn them in. It got pretty ridiculous!

Just the other day I was watching a woman who said that for the coming year of 1997 her show would do this and that -- obviously an old show!

Who Do You Know?

One guy came in and demanded that I air his old shows. After I said sorry, the department supervisor told me to air them. As you can guess this caused many problems for the traffic person, especially since the other staff members were doing favors for people too.

If some of this sounds strange, it is just the way the system worked while I was there, and I'm sure it still goes on at a lot of stations. People can't resist getting over on the system and giving out favors, it's human nature. Gifts during the year end holidays went a long way.

I mainly want you to have some insight into how these places can work. They all don't work like this, but some do and you have to be prepared for it. Above all, be honest -- especially with yourself and keep the good karma coming. As John Lennon said, "Instant Karma's gonna get you."

Your Debut

Okay, you have your air time and you call up your friends and family -- no network executives yet -- and tell them to watch. When they ask you what channel, you first ask them what cable system they have. Then you can tell them what channel the Public Access station is.   Every system has a different channel. If the people don't know what system they have, they can check their bill, and/or call the cable company -- like the ones in the back of this book.

ON LOCATION

Dollies' Dilemma; Celebrities in Cement; Driveways of the Rich and Famous; Rock in a Hard Place, Mistress Julie; Everyone Can Exercise; L.A. Nitelife; AIDS Walk; Faces of Hope.

Maybe you don't want to go into a studio to tape a show? Or maybe you need a video roll-in segment for your studio show? Well, my friend, this is the chapter that will teach you how to do just that. Read on, we're gonna have some fun.

I'm Your Captain

Would you believe I got to drive the "Goodyear Blimp?" Yep, I sure did! All because I helped an intern/producer shoot her show about the blimp. We were taping a segment with one of the pilots, up in the sky, and he asked her (the host) if she wanted to take the controls. She got in there and I taped her flying the blimp. Then he asked me if I wanted to try it. What do you think I did? That's right -- I jumped in there and took the wheel! It was very cool.

How about you doing a show from a big balloon? Or on a boat? Read on.

The Equipment

Many of the studios offer remote equipment: camera, 3/4" deck, tripod, monitor (very important), headphones, lights (two or three), mics, cables, batteries, etc., which you can use to record your show. They will also train you for FREE. I shot shows myself, and trained people for five years, how to shoot on location. When I started, I didn't have a clue how to use it.

Off To Vegas

Maria Serrao, who I mentioned earlier as the girl in the wheelchair with her exercise show, *Everyone Can Exercise*, asked me to go to Las Vegas to tape her show in a hotel workout room. The whole trip was FREE (plane ride, accommodations, meals, etc.), plus I made $100. We met impersonators (Cher and Elvis) and male exotic dancers called the "Thunder from Down Under" (Maria fell in love). We also met a few of the *American Gladiators* -- remember them and that wild show?

The experience was great. I went back to Vegas with her and got to see *Sigfried and Roy, Cirque du Soleil* and *Starlight Express*, all for FREE! She knew how to cut a deal. Remember Chapter 10

172

("Sponsors")? That's who paid for everything. Have I got you thinking?

I should mention that I shot it on my own equipment: Hi-8 camera, a couple of mics and a few simple lights (the flood kind from Target -- cost about $35 for both). But you can use the studio's equipment and get a great looking show from anywhere.

CAUTION: some studios don't have their equipment insured for out of state use -- check to make sure. I know a guy who was a pilot and flew the equipment to his location, then told us about it. We had to tell him not to do it again. If it got lost, he'd have to replace it; funny thing though, if you lost the equipment anywhere you'd have to replace it, didn't matter if it was local or out of state.

Training Tips

What about you? The studio will train you on how to operate their camera, so I won't go into all the details. However there are some keys things to remember. Pay attention to the "white balance" instructions. You have different settings (on the "filter wheel") for indoor light, outdoor when it's sunny and outdoor when it's cloudy. This information can help you out since it is used for proper colors.

Briefly, you hold up a white piece of paper to the person's face you are taping, zoom all the way in until the white fills the viewfinder/screen and flip the white balance switch up -- it should have "AWC" above it. Then you'll see something like this in the viewfinder "AWC OK." (sometimes there's an "A" and "B" setting). Most cameras have an auto white

173

switch that sets it for you. But I don't trust them. Do it yourself and check your color monitor.

Want to have some fun with the white balance? Instead of holding up a white piece of paper, try using a color, like red or blue and see what it looks like. Is everything green, blue or reddish? Yes, that's right, it puts a tint on the video. I discovered this by accident when I mistakenly white balanced on a "Coke" machine's red color. But the class loved it! You see, the camera only sees what you tell it to see -- if you tell it that blue is white then it will adjust accordingly.

And don't forget to go back to proper white balance when you need to, because you can't fix this kind of color later. Always check your color monitor!

Also you'll want to "Black Balance" too. No need to do anything else except press the same button, but this time down to "ABC" and you'll see something like "ABC OK."

Focus

Since these cameras don't have auto focus, "Rack Focus" is what we did: zoom in all the way to the person, then set your focus, and all your shots will be in focus, as long as the distance doesn't change between the person and the camera. Practice it a bit. And check that monitor!

Everything I learned, came from a lot of mistakes. Don't' be afraid to make 'em!

Shooting The Interview

You only have one camera, so if you're shooting an interview show, then try doing it this way:

1) Tape all your guest's answers first, while you ask them questions off camera.
2) Then they leave, and you have the camera person tape you asking the same questions to an empty chair, along with some reaction shots of you.
3) If you have time, and the guest can stay, have the camera person set up for a "two shot" or "set shot" that includes both of you (the "third" camera angle).

This technique is called "Single Camera" or "Film Style" shooting. When they shoot movies, they'll shoot the same scene over and over from different angles (points of view), while varying the shots from a master, to medium to a close up to extreme close ups.

Recently I worked on two movies of the week and a lot of the stationary dialogue scenes were shot this exact way. I'd say, "it all comes down to a simple Public Access talk show" -- the director and editor would look at me like I was crazy.

While you're taping, have the camera operator vary the shots. But keep it mostly medium, with a few over the shoulder shots (part of you from behind -- they'll know) and some close up shots.

If time allows, tape a segment at the start and look at it back on the monitor -- just to be sure the camera really is recording something, and to see how you like the angles.

Can You Hear Okay?

Don't forget the audio! Make sure it isn't peaking too much into the red area (on the "vu" meters), yet strong enough for a good level -- hitting "0" is okay. Have the camera person wear headphones to check for clarity of sound. Record a minute of the segment -- then play it back to make sure it is recording correctly. I can't tell you how many times a producer brought a tape back with nothing on it, or really bad audio because they didn't check it out first. You can do it the night before you go to the location, or better yet, when you pick up the equipment, that way if it doesn't work you can get one that does.

Uh-Oh It's Over $50!

Century charged $50 per day or weekend for the equipment. Well there goes the budget right? Wrong. Get together with another producer, rent it for the weekend and split the cost. Buy three used tapes (Chapter 8) for about $5 apiece -- let's see, that's $25 for the camera, and $15 for the tapes, a total of $40, and give the intern $10 for shooting it. Always give people something -- gas, food -- it doesn't matter, no one likes to work for free, whether they say it or not. Most of the places don't charge for equipment, depending on what you use (Century's Super VHS camera was free of charge at the Eagle Rock office).

Other Options

Or you can use a S-VHS, Hi-8, or Digital Video camera, then transfer it to 3/4" tape in the edit bay.

176

You have to buy an hour 3/4" tape for transferring. If you use another format, try to get the highest quality you can. The regular VHS tapes don't hold up well for transferring and repeated edits. You're better off using the Hi-8 S-HS or Digital. Also, make sure you still have the camera to take into the edit bay and transfer it to 3/4", then you can edit from it (next chapter). Patience, we'll get there.

On The Street

The 3/4" equipment is big and bulky, but they have batteries so you can use it for "person on the street" style interviews. Where I currently live, we have the "Third Street Promenade" which is very popular for these segments. I just did a walking interview with a New York based musician-singer-songwriter, Greg Kroll. They're fun to do and provide an interesting setting. And the police don't bother you either. We always see people down there taping something. Do you know of any cool locations where you could tape your show? The setting can make a big difference.

Please Release!

Don't forget those release forms! It will come back to haunt you. A "Release Form," says that you can use any portion of the interview for any purpose, without compensation -- all studios have them. TV News generally doesn't need these since it is news. My wife, my daughter and I were just interviewed on the "Promenade" for KCAL news (a Halloween story) and weren't asked to sign a release. But you do need releases, so bring plenty along and get them to sign them. I can't tell you the horror stories

I've heard about people refusing to allow their interview to be used, especially when it's really good stuff!

Okay, let's talk about some shows, and try to get you thinking more about what kind of TV show you're going to be starring in.

Let's Go Diving

A friend of mine, Nigel, did a couple "On Location" shows with a producer who liked water. Since he was a certified diver, they went out on a boat and taped a show underwater in a giant kelp forest that was beautiful. Then they did a beach report show and added a few surfing lessons. Nigel did have his own Hi-8 camera, which he charged the producer for using, but got a diving trip.

Do you like to dive? Public Access might not be the best equipment for that, but check with the studio and they should know of someone who has their own equipment.

Equipment Reality

A few more things about remote equipment. Depending on the studio, the equipment might be older and abused or newer and well maintained. Unfortunately Century's fell into the older category, but we were able to make it work. As I mentioned you could also use a high end consumer camcorder: Mini-DV (Digital), Hi-8 and Super VHS work better than other consumer types. Some studios also have matching editing bays too; for example, the Century studio in Eagle Rock has Super VHS editing, and the TCI studio in Van Nuys has Hi-8 editing. But they all broadcast on 3/4" tape. Later, when your budget allows it and you need a smaller camera to take on location, check around town for camera rentals. You can get a good one for about $100 a day -- the studio should know where they are -- in L.A.: Samy's, Bel-Air Camera and Alan Gordon are popular places to rent from. Otherwise use what they have and make the best of it. Remember, it's not the quality of the equipment so much; but rather, what you can make it do for your show!

Dooumentary

The studio equipment is good for doing documentaries too. I shot a documentary about Kelly and Sharon Stone's non-profit organization, Planet Hope, that helps homeless people find work, by training them on computers.

They also have a Camp Planet Hope program that sends homeless families to camp for a week. The kids have fun, swimming, singing, exercising, playing, etc. The parents learn skills of job interviewing, dressing up, and they get complete makeovers by volunteer stylists. Plus they get physical checkups, and follow ups. It's a great organization that was founded by the two sisters Kelly and Sharon (yes, she's the Oscar nominated actress).

We had a blast at their camp one summer, with Martin Sheen (yes, the actor) and his son Emilio, also came out for the day. The biggest thrill was talking to the people who were being helped by this organization, especially the kids.

What cause can you do a show about? During my time at Public Access, I was privileged to work with some other great organizations; including, AIDS Project Los Angeles (AIDS Walk; Dance-a-Thon, Commitment to Life, Back Lot Blowout); Mothers Against Sexual Abuse (who nspired our documentary, for which my wife, mother-in-law and I received Emmy nominations); PAWS; Soberlink; Battered Women's Shelters; Project Linus; My Friend's Place; Children of the Night; and Chrysalis to name a few. Pick a non-profit organization and do a documentary. Get committed and help out. Pick up the phone today!

It's All In The Lighting

Another factor about "On Location" shooting is to get a good lighting person (in addition to a good camera person). There are competent interns who can do this. But ask around and see if you can afford to pay someone. A friend of mine was charging $200 a day to shoot and light. This is something to consider as your budget and popularity grows.

I saw a lot of shows that could've looked great if the lighting was right. Here are some tips:
1) Make sure you have a color monitor to see what you're shooting. (Don't forget to white balance properly.)
2) Have enough light to see the image -- under lit looks orange.
3) Don't shoot in front of a white wall or with a bright light behind the person (they'll be silhouetted and you won't be able to see their face).
4) Bring some kind of a reflector -- those silvery, shiny car shades work well. Or buy some white

poster board, and on the other side tape some tin
foil. One side is soft light, the other is bright.
5) Put a blue gel over the artificial lights when
shooting in daylight.

To Tripod Or Not?

One last tip: use a tripod. Even though MTV started
a whip-pan-frenzy-herky-jerky look, if you don't do it
right, it will look bad. Start with the basics, then get
creative. I saw many shows come back to be edited
and you couldn't use much because they were all
over the place. You can do a wild show with
movement, but learn the basics first.

Also, another thing to think of is movement. Use a
"wheelchair" for a dolly shot. You can buy them
used -- look in the classified sections under
"Medical Equipment." Some people have even
used shopping carts for dollys or if you're indoors, a
rolling office chair -- I did that in my kitchen and it
looked great!

Remember Martha Quinn?

The cool thing about me moving on from Public
Access is that I get to share my new experiences
with you. For example, I just finished some non-
linear editing on The Martha Quinn Show -- she was
one of the original MTV VJ's (TRIVIA TIME: Can
you name the other four?). The segment called "In
the Zone," featured Martha and her friend, Wendy,
exploring places around Los Angeles. They did
everything single camera -- constantly moving
between the two girls. It worked because the show
was about having fun and the fluid camera style fit

perfectly. Plus they had a lot of "B-Roll" shots (i.e., they mentioned a drink and you cut away to a person making a drink, or a book and you see a close-up of it or talk about a massage and you see a massage).

Get plenty of "B-Roll" shots so you can do cut-a-ways; that is, if you have that kind of a show.

Studio On Wheels

Some studios offer a fully loaded production van to go on location and tape your show. I've never used one and didn't know anyone who did. But I've heard that it was a lot of work to organize and you had to fill out paperwork, plus take out a hefty insurance policy. However, I encourage you to inquire about renting the van.

Start Simple

But for now, let's use the equipment they have for on location shooting, get comfortable, do a simple one camera show (in your house or back yard), then work your way up from there. You will have an incredible amount of fun learning and making mistakes. I want you to tell me about it -- hey, send me copy of your show, my address is listed in the front of this book.

Now let's take a look at what you do with that footage you shot on location. Be brave, we're going into the world of editing. Seriously, it's a lot of fun!

182

EDITING

Perhaps you've heard of the movies, *El Mariachi*, *Desperado* or *From Dusk Till Dawn*. They were all directed by Robert Rodriguez, who edited his first movie, *El Mariachi* at a Public Access studio in Texas, using their 3/4" edit system, just like the one you can use. Maybe you could edit a movie on it too, someday!

Don't Be Afraid

After you shoot your show on location, or want to make a "Roll-in" reel, you go to an editing bay or editing suite (fancy name). Most Public Access facilities have at least two edit bays -- some are more advanced than others. No matter how fancy the machines are, editing breaks down to just a few basic fundamentals. Remember, I taught editing to people for five years -- even those who were afraid of machines. They all learned it in a few hours -- so don't let it scare you.

Here's my simple approach. Editing bays have two machines -- one for playing and one for recording. What you want to do is take your play footage and transfer it in a certain order to the record machine. Easy enough you say? Okay, now let's learn how.

Vinyl Flashback

Remember the days of making your own cassette tapes from vinyl albums? (Ah-ha, caught you on your age, didn't I?) You might have liked a few David Bowie songs, a few Alice Cooper songs, a few Led Zeppelin songs, throw in a Ramones' song and a few from Earth, Wind and Fire or Parliament Funkadelic, and there you'd have a custom cassette. It was easy, all you had to do is cue up the needle to the beginning of the song, push play and record on the cassette deck and it would record it. Then you pause the cassette deck at the end of the song, cue up the next song, release pause and so on.

Well that's all you're doing here too -- putting a certain amount of video tape in order on another tape -- known as the master tape, which will air on Public Access.

Enter The Bay

Now you go into the edit bay, put your play tape in the play deck (left side) and your record tape in the record deck (right side). Cue up your play tape to a point where you want to start it playing and mark it as the "IN point" on the edit controller -- which sits in between the two decks (the most popular edit

184

controller is usually a SONY 450 or 440). You mark
it by pressing the "ENTRY" and "IN" buttons
together. Then you cue the play tape to the out
point -- where it will stop and mark it the same way,
by pressing "ENTRY" and "OUT" buttons together.
Now go to the record side and cue the tape where
you want your play segment to begin recording.
Mark that as the "IN" point, the same way (did you
notice the "IN" and "OUT" lights stop flashing and
stay red?). Press "PREVIEW" to look at it before it
records. If you like it then press "RECORD." You
repeat these basic fundamental steps over and over
again, until you are finished.

Step Back To Pre-Black

There is a minor thing you should do before you
start to edit, and that is to pre-black your record
tape before you start editing on it. First make sure
you have some kind of "Black" source (usually it's a
button on a machine marked "FADE TO BLACK") --
just make sure the record monitor is black -- (not
fuzz or anything else, just black). Put your master
tape in the record deck, rewind it to the beginning,
press the "ASSEMBLE" button (a red light goes on),
reset the tape counter, then press "PLAY" and
"RECORD" buttons together, and let it record black
for the entire tape. Also make sure the numbers on
the record deck are counting/moving -- that will tell
you that there is a control track recording on the
tape.

When it is done, it will automatically rewind to the
beginning. Press the "ASSEMBLE" button again so
the light goes off, and then press the "VIDEO" and
"AUDIO 2" buttons (under the "INSERT" heading).
You're telling the machine to record the video, and

audio on track two only. You'll also want to start your record tape at about 1:00 into the tape (play or search ahead -- press the "Search" button and grab the big black knob) -- that leaves you room for "Bars and Tone," which is a video and audio reference for playback. They should have "Bars and Tone" on a separate tape -- ask for it. You'll later record 30 seconds in the middle of that beginning blank 1:00 space. Your tape should start with 10 seconds of black, then thirty seconds of "Bars and Tone" and then 10 more seconds of black -- you've left ten more seconds of black, just in case you need a little extra.

You might want to leave another ten or twenty seconds of black for an opening which you can create later and come back to put it in. Just remember to zero out the counter where you want your show to begin -- you only have a thirty minute tape and need to know how long the show is while editing it. Remember, you need time for your opening and closing credits too. Yes, you'll have to type them in yourself -- the instructor can show you how, or get an intern to help with them.

Something Fanoy

Here's how I created a fancy montage (pictures and audio) opening for my first show. I rented the camera and used it to record all these pictures of musicians and bands from any source I could find: albums, CDS', cassettes, magazines posters, etc. I recorded about ten seconds of each picture (about a hundred or so) on a 3/4" tape. Then I got this fast paced music -- I made it myself on a home four-track recorder using a drum machine and synthesizer -- to use as my theme.

186

When I went into the edit bay, I first recorded the music on the "AUDIO ONE" track only (audio two was normally used for dialogue)for about thirty seconds. Then I listened to the beat in the "SEARCH" mode (using the big black knob -- frame by frame), which you can control, to find out where each picture would change according to the drum beat (snare hit). It was about ten frames for each picture. So I had to record ten frames only of each picture, for thirty seconds. Which if you figure it out, each second has thirty frames, and I had three pictures per second . . .which figures out to be ninety edits (pictures). Whew! It was work, but looked great when it was done.

On With The Show

Now you're ready to start your show -- let's try editing an interview. You find your interview footage of yourself asking your guest a question. You mark it and record it. Then you go to your guest's answer (on the same tape or did you use a separate one?) to that question. Mark it and record it. Go back to your next question, mark it and record it. Go to your guest's next answer, mark it and record it, etc.

If you used separate tapes it might be easier, but remember to keep resetting the tape counter and rewinding the tape to the beginning, so you know where to find the segments that you logged already.

A Long Answer?

How about a quick "Reaction Shot?" Remember when you were sitting there nodding to an empty chair? Stick it in for about two or three seconds while your guest is talking. But remember to only use the "VIDEO" channel -- turn off the "AUDIO" record channel (the red lights goes off), when you edit. If you leave the audio record channel on, you'll delete what your guest is saying -- don't do that! If you always "preview" your edits, then you'll know if there is a problem.   Also make sure there is enough room before and after your reaction -- at least five seconds to avoid what would look like a mistake -- a jump cut.

Did You Log?

At this point I should tell you it was a good thing that you had someone keep track of where your questions were and where the guest's answers were, time wise on the tape. Even if you didn't do it on location, you can do it while your record tape is pre-blacking. Having that list (tape log) of where everything is will save you a lot of time when your going back and forth between tapes. Log those segments, please!

Keep It Simple

I won't get into graphics or special effects here -- the studio should be able to show you that stuff. But know that it is very easy to edit your show, and you don't need alot of fancy graphics or effects when you have a great show with a great guest! Sometimes those fancy effects take away from your show -- use them sparingly.

Advanced Idea

Something advanced? How about a montage opening of your show after you make about four or five shows? Choose your best moments and put them down to the music -- just like I did with those ten frame edits. You don't have to put them in so fast, just as long as they go with the beat of the music. I edited some music videos and movie trailers like that and it always worked when the music guided the images.

A Few Notes

- Always check your edits as you do them.
- Make sure your audio levels are strong enough (close to the "0").
- Once you've pre-blacked your tape, never go back to "assemble" editing.
- Ask for help. Some interns are pretty good editors.

Other than credits you now know how to edit your show. And remember to leave about a minute of black at the end of the show -- counting from where you fade to black.

Okay, you've got the editing down, how about learning how to work in an actual studio?

P.S.

I'm currently training people how to use a non-linear editing system (Discreet Logic's Edit*) which is amazing. When you get comfortable with editing, I encourage you to check it out. They're on the web at "discreet.com" or write to me.

You can edit faster and be more creative. I'm currently editing a cable TV movie and its a blast! Plus, I've worked on movie trailers, documentaries, music videos and commercials, too. And I get to work with veteran editors and learn from them.

I just finished using the edit system on the movie, *Flintstones in Viva Rock Vegas* as the on-set compositer. Watch for my name when it comes out in May, 2000. It'll be at the end, and probably very small.

INTERNING

You're over 30 and work full time. Why would you want to intern three nights a week or all day Saturday? Because you will learn how a show is put together, and the more you learn behind the scenes, the better your show can be. Even if you don't want to work in a TV studio, it's good to know how it all comes together. Plus you'll get to know the people and get their input about what works and what doesn't for a show. You might even ask them for some ideas on how to improve your show.

When I Was An Intern

I started interning in 1991, five nights a week, before I taped my first show. I got to know the people who worked there and felt more comfortable doing my show. The other interns all helped out and gave me a lot of encouragement.

For my first show as an intern, I was put on camera (wearing headphones) and they gave me some simple instructions -- "pan is left and right, zoom is in and out and tilt is up and down, this thing is for focus and don't move until the director tells you to."

Also, they told me to give cues, but I wasn't told what they were, until the time came for them. Here's some of what I heard on the headset during my first few weeks: "Okay camera three (that was me -- I learned the hard way) go get that picture." "Camera three are you awake out there!" "Hello camera three." "Who is that on camera three?" "Now camera three zoom out quickly." "C'mon get it together!" "What's wrong with you?" "Pan left, no the other left!" "Go, go go!" "C'mon focus -- what do you need, glasses?"

A little embarrassed? You bet! (I'm laughing now as I read this.) But the other interns were very comforting as they told me that some of the directors were real screamers, while the other directors were more calm and pleasant.

No Spielberg or Sayles

Over the first year, I got to work with all the directors and found they had unique styles and personalities. One was a comedian, who would make you laugh every second. He said, "it was his job, to make you laugh throughout the show." Most of the time, it was more entertaining in the control room than the studio. I laughed so hard during one show, the producer made them stop and remove me from the camera. The only bad thing was that this "funny" director was sexual with his jokes and caused some discomfort with the female interns. I understand he's directing at a sports channel and still telling those jokes. He was always the life of the party and interns loved working with him.

Can You Hear Me?

Another director was a screamer -- the saying was, "if you survived headsets with him, you'd survive anything!" (I understand he still screams at camera people where he works today). You see, everyone wears headsets to hear the director -- so they all know what's being said and who is being reamed. Believe me everyone got yelled at by him. He was generally a nice guy, but when it came to show time, look out! The good thing was that all the interns bonded together and were very supportive, so no one felt bad when they got yelled at.

More Screamers?

My wife worked with a couple movie directors who were also screamers. And I must admit, when I got promoted to director, I went through a period of the same type of "aggressive" behavior, but luckily was consciously shown how it affected people, especially when I took over the intern program and the interns would come to me to complain about other directors treating them harshly. I saw how it wasn't necessary to yell to get something done. Sometimes in the heat of directing you can lose sight of how people feel, but this is not an excuse.

And the main problem was that most interns didn't know anything about how to professionally operate a camera -- they were there to learn, but the directors expected a lot more from them. And we were going through so many interns that they didn't have time to get to know each other.

193

I hope I'm not scaring you out of interning -- what I'm really telling you is the worst it could be and will probably never be like this for you, but if it is, leave and find another studio, you don't need to listen to some raving-insecure-low-self-esteemed-maniac!

Referee

I ended up being in the middle of many disputes, but was able to resolve 95% of the problems. I only had to let two interns go in five years. One was a mistake, I sided with a fellow director (she had an ego problem), but shouldn't have, and the other was a guy who just didn't fit in anywhere. You can always resolve differences, and if it gets to an extreme, just chose not to work with certain people -- it's your right.

Contests and Parties

Okay, so you know how bad it can be -- which it really isn't for the most part. One of the best parts of interning was when we had the "Holiday Video Promo Contest" -- the first year I was there we (the night crew) won it. I wrote and rapped a special holiday song, and we acted out the lyrics in the studio about what we did. They went something like this: "We're the interns and we say, Merry Christmas and happy holidays . . . twelve weeks in a row, at the Public Access studio, we set plants, set lights, move chairs and put on mics. We're the interns and we say, Merry Christmas and happy holidays."

We had a blast! It was a fun time to be an intern and part of a team. There were some very creative promos coming from the other interns too. One team had their faces super-imposed as Christmas balls on a tree and they talked to each other. Then another used the time to be an "Intern News" show, featuring various holiday theme stories -- like a singing Christmas tree; a woman who never ages and is seen throughout history in an Egyptian painting, a Woody Allen movie, and the film, *Miracle on 34th Street.* They were very creative and used the Chroma Key effect a lot.

And I made sure we continued doing them when I took over the intern program for the following five years. It actually made me proud to encourage people to come up with creative ideas.

In addition, we would also have a big party in the studio during the year end holidays. It was the time to celebrate the interns who were working for free. Eventually it made it's way to an outside location -- the Santa Monica Pier in the Carousel room -- yes I said a carousel, and we had it catered, too! Friends and family were invited to come and watch the "Holiday Promos," then we'd give out prizes for the best in different categories -- actually everyone won a prize, then we'd have the grand prize.

Intern Of The Year

Also at the party, we'd announce the "Intern of the Year." They were picked from the eleven (we were always one behind) who won it each month that year. Each intern of the month got a pin and certificate. The certificate was created by another intern, Allan, a graphic artist (who illustrated this book). They'd get to make a speech, and we'd put their picture up in the control room -- it was really fun!

From Intern To Director

Interning will give you a great atmosphere to learn in, have fun and meet some very interesting people. And when it came time to hiring a new director -- it almost always came down to an intern. I was hired not because I knew how to direct -- I didn't direct a show, ever, but because I put my year in as an intern and could work well with people. They knew me and liked me. That was it! Everything I learned I learned on the job.

Playboy Bunnies & Exotic Dancers

Funny thing about Public Access was that no one really started out working in television. It was like a second career place. Some of the staff's former jobs were: Playboy bunny, bank teller, exotic dancer, nurse, financial specialist, Disney costumed characters (Goofy and Tigger), janitorial worker/owner and computer specialist. As mentioned earlier, I myself worked in *Billboard* magazine's mailroom. I also had my own business which was creating cool cakes for rock stars, back in Philly.

Even the interns were a diverse group: artists, musicians, lawyers, doctors, actors, salespeople, TV directors' kids, college and high school students, etc. No one came from the same background, which made it very interesting and good for learning.

196

From Intern To E!

My wife and I were flipping around the TV and
stumbled onto the E! Channel's *Gossip Show*.
Guess who we saw? A former intern named Laurie.
Just a short time ago she was an intern, then
started her own music video show, and now she's
on E!. Pretty good. I am never amazed at where
people can end up after working at Public Access --
it truly is a training ground for talent (besides I met
me wife there -- and proposed to her on air, but
that's another story -- what? You want to hear it?).

Yo Heidi

Okay, I was teaching my ENG class, in Septrmber
of 1993, and she, along with her mother, were in
this class. I thought that was cool. Her mom was
very funny and creative. They gave me a "thank
you" card at the end of class, along with a toy ray
gun. So I asked Heidi to be an intern.

We became friends from just working one day a
week together -- Friday. We scouted locations for
the "Friday Holiday Promo" -- (it was at the beach).
We went to lunch, saw a few movies, went bowling,
etc. And one guy came up to us on the Third Street
Promenade and said, "Did you marry her yet? I
said, "No" and he said, "There's something wrong
with you."

Then I found out for Christmas, she was going to
New York to visit her Dad and I was going to Philly

to visit my Mom. Let's just say we fell in love in the Big Apple, on the coldest day of the year!

When I returned to L.A. I started working on a song called "Yo Heidi" -- I wrote and rapped (not too bad for a white guy) -- with lyrics like, *I met your mom and your dad/when we're apart I feel sad/Yo Heidi, Yo Heidi* -- then an intern, Henry Harmon, would say "What's Up, What's Up" (he gave the rap song some authenticity) and I'd say, "Yo Yo Yo Heidi," and he'd say, "What's Up What's up What's Up What's Up."

And the best part was, we all got together and shot the music video. The interns all goofed around and it looked wild! They were all singing the chorus, bouncing up and down dancing -- you had to see it!

Then I put it into a show where I called people (friends, other producers, family members) to ask their advice on how to propose and people would give advice on the air.   And I took some clips from her favorite movies and put them in there too.

Then it aired on Valentine's Day 1994 -- a Monday. The weekend before, we had taken a romantic trip to a nearby mountain resort. We rushed back to watch the broadcast. Of course, she didn't know it was on and I literally had to grab her out of the bathroom to see it on time.

We watched, she laughed, and at the end she looked at me and said, "What are you trying to say?" I knew she was the right one for me!   So I asked her there in person after the show and she said, "When do you need to know by?"

Put it this way, the show was a success, we got married on June 18, 1994, which is Paul

198

McCartney's birthday -- and did you know he got married on mine -- March 12th? And now I'm the proud father of a daughter born on January 27, 1997. And it all started at Public Access!

Let Me Know

So check out interning, you might like it and get hooked into working at a studio -- maybe even find a lover and friend. All you have to do is make a commitment and try it. Even if you do it for just a few weeks you will have gained some insight into how the shows are put together. You can contact me with your behind the scenes stories -- I'd love to hear them, and maybe include them in my next book.

Don't Go There

One quick note though, some studios are very militant with their intern programs -- one of my former interns told me he hated going to another studio because it wasn't fun. I told him, like all the other interns, "When you stop learning and/or having fun, it is time to leave." There is no sense in staying at a place that is miserable. Your time is precious and people should appreciate your work.

Two more chapters to go . . .

LIVE SHOWS

The Boom Boom Room with Ray Mancini; Horoscopes by Larry; Stargazing; Law and You; Unzipped; Shut Up and Eat; Hollywood Yogi; Ask Mr. Traffic; Kramer's Corner; The Feelies; Dre's World; Jobless Joes; Ask Livia Live; Rudy the Rubber; Shirley U Jest; Love Works.

So far we've been talking mainly about "live-to-tape" shows -- where, if you make a mistake, you can stop the tape and either start over, or pick up where you left off. Where I worked, ninety-five percent of the shows were like this -- they left the studio finished and didn't require any editing. However, there is another kind of show that you don't have to tape at all -- although you might want a tape for your collection.

On A Saturday Afternoon

When you're confident enough handling weirdos calling in on a Saturday afternoon, try a "Live" show. You will probably have to wait about six months to get one -- they book that far in advance -- but the experience is worth it.

Here's how they worked. Every Saturday afternoon, we would have three "live" shows, after the morning "live-to-tape" show from 9:30 to 11:30. The producer would come in at 11:30 -- hopefully the morning show was finished -- if not, there'd be a mad scramble to set up the first live show -- and we'd set up and be ready at twelve noon for the first "live" show to begin.

They're Calling

The show would start like any other, and we'd put up a phone number for people to call. You had the option of taking calls two ways -- screened and unsscreened. The first way was one our interns would screen the calls for you -- they would ask the caller their name and phone number (and sometime what their question was), then hang up, call them back and place them on hold. You would see the light flashing on your phone in the studio and know there was a call. Plus we would signal you with a sign to let you know you had a call. You had a maximum of three lines coming in. It was up to you when to push the button and speak to them on the speaker phone.

We had a host actually pick up the phone and start speaking to the person, which was bad because the little microphone used to hear the caller was on the telephone's speaker, and no one could hear the caller.

Oh, and another thing, if a host wasn't used to speaking to a caller on air, they'd always be speaking directly into the speaker phone (like you would normally do at home) -- which of course looked like they were talking to the phone.

So two things: one, don't pick up the receiver; and two, don't look at the phone, look at the camera when speaking to the caller/audience.

They're Goofing

Sometimes we'd call back and it would be the wrong number. Or sometimes the people calling in would hang up when we'd ask for their phone number or just start cursing right away (thinking they were on the air) -- believe me, there are more "phony phone callers" out there than people think. From listening to "Shock Jocks" like Howard Stern in Philadelphia, I've learned a lot about people who have nothing better to do than make those calls. It can be frustrating to work like that, but you go with it.

Even though we would screen the calls, you would still get those crank calls coming in. If people were determined, then they would get on the air and try to make the host or guest look like a fool. And if the host wasn't ready, they'd be dumbfounded. One host was talking about sex -- he was a doctor who specialized in penile enlargement -- and people were goofing on him the whole show. One caller asked him, "What did you do with the money?" To which the host replied, "What money?" (This I knew was a big set up). The caller replied back, "The money you saved on your cheap toupee!" Everyone laughed at that one! The host was flustered.

Hey, Baby

Even some women hosts would get those guys calling in and asking for a date. Or even worse, they would say sexual acts they'd like to perform on the host -- quickly the host would scramble to hang up. It was sad to hear, but I did realize how many sexually perverted and scary people there were (and still are) out there. My heart went out for the hosts on many occasions, but there was nothing I could do, we would screen the calls as best as possible and pull down the phone's mic if we sensed a crank coming. I even called people back and told them how wrong they were, but that seemed to fuel their sickness.

No Screening

The second way to receive the live phone calls would be to take them unscreened -- using the phone we used in the control room, in the studio. In the beginning we had a problem setting up, because the phone extension wasn't long enough to reach into the studio, and we couldn't turn the ringer off -- it was actually funny. But we figured it out.

Good hosts could take the calls as fast as they wanted to. They'd push a button and the caller would be on the air. Although the host would have complete control, they'd run the danger of more goof-offs! Unscreened calls were usually reserved for the more experienced hosts, who could turn the crank call into jokes or shoot right back -- like a stand up comedian handling a heckler.

These Work

Shut Up and Eat, was situated in a diner -- an interior picture of an actual diner superimposed on a blue screen (there's that Chroma Key again) background -- with a waitress, menus, donuts, etc. The two hosts would take calls and talk about current events. It was on during the O.J. Simpson and Mendenez trials and got a lot of good views and opinions. Funny, there were hardly any phony phone calls. They had intelligent people calling in to talk about their opinions.

Horoscopes by Larry, who was an occasional guest of the popular Los Angeles morning radio show, *KROQ's Kevin and Bean Show*, got a lot of good calls. Although the DJ's constantly made fun of him on their radio show, people calling his "live" show would take him very seriously. You would tell him your birth date, and time of birth. He'd look the information up in a special book that told him your astrological sign, rising sign, etc. Then he'd tell you a little bit about what was in your future and how to deal with certain situations.

Anne Shaw and her *Stargazing* show was also popular with callers on Saturday afternoon. She was good at entertaining people, by being tuned into what was going on in their lives. But please don't call her a "psychic," her shows were based more on astrology, and she had this sense about her that you can't describe, you just had to see her. Besides that, she was always fun to work with.

Ask Mr. Traffic always got non-stop calls. People would ask him things like, how to beat a ticket, who had the cheapest insurance, rules of the road, and even silly things like, can a guy beep at a girl walking down the street -- the answer was "no." He would also update you on the latest in DMV state laws and give good advice on how to stay safe on the road. He became very well-known around Los Angeles.

Kramer's Corner didn't give you information about traffic laws, but Mark was a fun guy to call in and talk to. While he was mainly talking about sports -- a very popular subject for "live" shows -- he'd talk about anything the callers wanted to. This former New Yorker was also an aspiring actor, and promoted himself this way, as a character type. He got a crank call once in awhile, but would end up making fun of the person calling in, "What are you doing in on a Saturday watching me, don't you have anything better to do?" We always had fun with Mark, and the show for the most part always went well.

Are you thinking of a "live" show? I talked about Astrology, Sports, Traffic, and a Diner -- what is your show going to be about? Do you think you are ready to handle some phony phone callers? Are you excited about talking to people "live" on the air? Are you prepared to do a show if no one called in? Yes, that did happen sometimes.

Music Videos

Remember earlier I talked about *Dre's World*, which featured soul, rap and R&B music videos? A former intern, Andre, took requests from callers to play music videos. This was a very popular live show and the funny thing was that he only had a limited number of videos and people would usually request the ones he didn't have, since he had no way of knowing ahead of time what people wanted (like they do on the radio). But he got non-stop calls and had fun!

His catch phrase was, "Represent Your City." When a caller called in, they'd say, "Mike from Compton," or "Melissa from West L.A." People called in and never made a crank call. I would be Captain G, the DJ, and interact with him. He'd say, "Captain G, you ready for the next video?" And I'd come on the intercom and say, "Yes, my brother." We had a ball! Sometimes we'd even take a camera into the control room and do part of the show from in there.

His show, like other intern-based shows grew out of boredom and cancellations -- people not showing up for their studio taping. So one day we just started playing music some of my music videos, Dre talked to people, and the rest is Public Access history.

Texas Style

From Austin, Texas came *Ask Livia Live*. Think of an attractive woman, with   an attitude, dressed in a tight dress, stretched out on a couch, talking to people about their relationship and sexual problems. Remember Dr. Ruth, the older-sexual-therapist-type lady? Well, she was a saint compared to Livia!   It was amazing how many people called in to get advice from her, and believe me she wasn't shy about giving her opinion either.

Behind the attitude, she was a really sweet lady who was always nice to the staff and crew -- never any problems -- but when she got on air, she became the character. The local news ran a story about her

206

-- during the day she worked as a scientist (I think she had her Ph.D.) -- then they'd show her doing her "live" show on Saturdays.

Dave

One show that wasn't sexual in nature, but was very complicated and innovative, was a show by this guy Dave. He had it worked out to where people could watch the show and interact with it on their computer. It was a first, but not repeated too much. I think it became too difficult to do.

Silly

Occasionally a producer would cancel at the last minute, or not show up -- in that case we'd make up our own shows. Most of the time it became the silliest thing you could see. Who ever was the scheduled director would get together with the interns and make up a show. One of my former fellow directors had a show called, *Jobless Joes*, where he, an intern, Justin and the Dave producer I just mentioned, would lay around and answer the phone, while people called in and goofed on them. They actually encouraged people to call and goof on them. For the most part, people loved it.

Talking Condom?

I was just by the old studio, on a Saturday, and ran into a producer who was going on to do a live show. His name is Michael and he's the voice behind the little smart aleck, Rudy the Rubber -- yes, he's a talking condom. Along with co-producer and Rudy's

207

sidekick Shane (a perfect low key straight man, who dressed in pajamas) -- these guys have put together a comical and helpful show. While they have crossed the line a few times (a topless dancer) for the most part they were and still are doing good work. It's fun and they have the involvement of APLA, who encourages people to have safe sex.

I directed their very first show at the studio just a few years ago, and they have since gone on to build a large fan base -- people are talking about the show on local radio. Rudy also got a lot of calls whenever he was on, and has had some very entertaining guests -- but Rudy was always the focus of attention and he ran the show.

Do you have a character you can create? A puppet of some sort? Plan it out and do a show around it, just as these people have done so far.

Telethon

When I first started working at Public Access, I took a "live" show and turned it into a telethon. The first ever attempted for Public Access. It was the usual half hour, and benefited APLA (AIDS Project Los Angeles). We tried it in July of 1992 for half-an-hour and it was a big success. Then we extended it to three hours in December of that year -- where we went "live" back and forth between the two studios, and it was an even bigger success.

Again, we didn't really know what we were doing, so I invited all the other producers to come down and join me for some fun. We called it *Glenn's Garage,* and it turned into a big dance party. We took phone calls, made challenges to the audience, raised

208

some money and most importantly, awareness. Mr. Traffic challenged people to meet and beat pledges and offered free advice on how to beat a ticket. Chris the Magician did some tricks, then his mom called in to pledge money and challenged other producers to do the same. We also had Suzan Stadner come on and demonstrate the correct way to apply a condom. People from APLA stopped by with information. Musicians, Starr Parodi, Auto & Cherokee and actor Dennis Cole came by to support the cause. It was a very proud time for us who worked there -- and it all started with an idea I had for a telethon.

Boom Boom

One last show I'd like to mention is one I did live with the former world champion boxer, Ray "Boom Boom" Mancini. He let me drag him into the studio for two live shows. We had worked together on a Pay-Per-View wrestling event in Dallas and on Alcatraz Island. To make a long story on how we met (I'll save it for my next book) short, I asked him to do a show, he said okay.

It was just Ray and a phone, and the callers loved it. He had fun giving his comments about current and previous fights. Some of his comments were, "it's obvious you don't know the difference between a fishing hook and a left hook . . ." and "You take a beating, that's what you do. . . . "

We kept in touch after the show. I created a pilot show for him that is currently in development, and we are looking forward to working together again. He's busy making movies and being with his wife and children.

Have Fun

So, if you're interested in doing a live show, pick a good subject and have fun with it. Be entertaining, informative and most of all, be outrageous. You can do it.

CONGRATULATIONS!
YOU'RE THE STAR!!

20

WHAT'S NEXT

You can do your show for as long as you like and are having fun! I mentioned other cities and you should be networking with other producers to find cities to air your show. You can also start a fan based mailing list and get people to maybe start coming to see your show as part of an audience.

Life after Public Access? Sure, it is a lot harder to do, but you can start running commercials and making some money by moving over to "Leased Access." What is it?

Leased Access

It is just like Public Access except you can be commercial. Which means you can make money -- at least enough to pay for the production costs.

I've seen a chiropractor have a show on Public Access and leased access at the same time. The difference was that he could advertise his costs, when he had hours and where his office was located, on the Leased Access Channel.

The same was true for a hair stylist who started on Public Access, then went on to make more of a commercial show -- it was one big commercial with his phone number always on the screen -- something you couldn't do with Public Access. It was basically the same show, but with more of a sales type approach -- always saying how good you can look from a haircut by him and how little it costs, etc. It was shot in his salon.

Free Pizza

What about a local business that needs to advertise? Let's say, your favorite pizza parlor. You could feature them on your show, get free pizza and make some money. How? Go to the restaurant and talk to the manager about your show and what it can do for their business. Do you have any viewer letters or reviews of your show? Any past shows that went very well and your guest can highly recommend you? Give them a VHS copy of your best work.

When the manager gets back to you -- asking how much it's going to cost -- you can tell them exactly how much.

Expenses

What does it cost? You have to rent a camera package, a person to run it, and a few extra crew members. You can get a good package for about $300 for the day. Then you have to edit it. Which

212

you could do at a local edit bay for about $25 an hour -- you'll need about eight hours (that's $200). In Los Angeles, there are edit bays listed in the "Recycler" and "Hollywood Reporter."

Then there is the cost of airing the show. It all depends when you want to air it, which means that the "prime time" spots can go for $1,000 per half hour and the "not prime time" slots can be as little as $100. It also depends on the cable company -- Century was charging a premium price for their time slots -- they had a great coverage area. United Artists (TCI), on the other hand wasn't charging nearly as much for their air time in the Valley.

Century also required a million-dollar insurance policy against liability -- which could cost about $500 to a $1,000. Even though I spoke to a rabbi who made the move to leased access and he told me he didn't have to pay for the insurance. You should call up a show you see and ask them about how they do it. Then call the cable company's leased access department.

From my experience you should get at least $2,000 for a show. It is a lot more work and you need some negotiating skills, plus since the "client" is paying you, you have to answer to them for the finished product. They might not like something and you have to change it -- costing more money.

If this isn't what you want to do, have them on as guests, or tape a short segment for your Public Access show and get some free pizza, and don't worry about what they think.

Special Effects

Let's talk about some simple special effects you can do to add to your Public Access show.

"Keying a Title." The producer/star from *S.O.A.P.* would bring in his logo -- black letters over white. Then we would use a key effect to either superimpose it over the set or change the black letters to a color, or the white background to a color. Ask the director to show you this technique. You can create a logo on your computer and then bring it in and have the director put in front of the camera, freeze it and manipulate it.

Speaking of a "freeze," it is a still image which can be stored electronically in a TBC. Then you can change the image with the poster, mosaic or sepia buttons (Note: video, chroma and setup levels can't be change after you freeze). Always have it backed up on tape, so you can recall it and change it.

What you can do is take a freeze of yourself, you and your guests, the set, a book, a picture, etc., and use it as your opening. You can also take multiple freezes and dissolve from one to another. This takes some time and requires at least two TBC's and a competent director. They have to record the two freezes in the TBC's, then record them on tape, then lose the first one while retaining the last one and so on. It can be confusing. But it looks good.

Camera Fun

Some camera techniques can be fun. Have the director set up camera two on the set shot. Then move camera one behind camera two to shoot its monitor. So you have a set shot on camera two and camera two's monitor on camera one. You have camera one pull out to reveal the camera and the set, then dissolve to camera three (your camera) for your opening.

Or even placing a camera on a slight angle, on a tripod, can be fun -- (like the old *Batman* TV show with Adam West and Burt Ward) don't get too wild, it takes away from the content of the show.

You and a Monitor

One of the hair stylists would do a show by himself and a monitor. He'd sit on the set with the monitor and use it to roll-in his video segments. The trick was to always have something on the monitor -- like his logo. As he would introduce the clip, we'd push into the TV and then dissolve to the actual clip playing on another tape deck. Then we'd cut back and forth between him on the set   It was tricky to do, but again, looked cool.

This "monitor on the set" technique also worked very well with music videos and movie clips shows. The director should know how to set it up -- just tell them what I told you, to route it so it is not the main record feed (you'll get feedback), but rather a TBC feed, which allows you to put any camera or tape deck through it.

Money

Some producers made a living out of what they did on their shows. All the hair stylists did. Some cooks had their own books out. Then there was this one woman, who did a popular psychic-reading live show. One day I saw her giving readings (and making money)outside a local food market, Wild Oats, in Santa Monica. She had been quite successful at making money from her show by giving personal readings.

And Anne Shaw was popular at parties with her stargazing readings. (Remember her from the previous chapter?)

Demo

How about a demo reel of your show? If you're looking for an agent, you can put together a highlight tape of your best work -- especially if you're an actor. Public Access is a great place to hone your skills, no matter what theme you're doing.

Also, some successful network producers would come to Public Access to work on new ideas for shows and not have to spend a lot of money.

News

News? If you're excited about covering those real news events, get a good demo together that shows you interviewing people on the streets. Show how you can cover some local events. One woman I know of got a job in Fresno as a reporter for a local network station -- I believe it was a Fox affiliate.

One producer/co-host, of a sports show, went on to a Georgia news station, to work as a weekend sports anchor, where he was actually shooting and

editing his own news segments. Both began on
Public Access.

Big World

Have you checked out the rest of the country, or the
rest of the world? People are always looking for
new and innovative shows to put on their channels.
Go to the library and look up small local channels.
Call them up and send your show to them.

There are also a group of Public Access support
organizations out there that can help you with airing
your show around the country, and even overseas --
check with other producers and the studio
supervisors, or the back of this book!

Movies

Did you know Movie Studios call Public Access all
the time for background TV footage? When you
see a movie with a TV on in the background, there's
usually a show on it. To get that footage, a studio
usually has to pay a lot of money to license it, but
Public Access people donate it for free. It's thrilling
to tell people their show is in a movie. My wife was
contacted by the producers of the Sally Field/Ed
Harris movie "Eye for an Eye."

I often see movie studios quoting Brian Sebastian
on his Access TV show, *Movie Reviews and More*,
for ads in local newspapers. One said, "Funniest
movie of the year."

City TV

Also, check out your local City TV for special projects. We went to Santa Monica's City TV and received a grant to use their Beta equipment. We produced a documentary, which received an Emmy nomination.

A Last Thought

A final word on what not to do. Don't use your show to be nasty to people, it doesn't help further the genre, or society. There was a show where two guys would come on and make fun of Jews, blacks, women and homosexuals in a mean-spirited way. It would upset a lot of people -- which they got off on. You don't have to cut people down to make yourself feel better -- if you hate yourself that much, get some help, or better yet, stay off TV.

You don't have to curse excessively to make a point. People really do understand without a lot of unnecessary adjectives and adverbs.

The idea is to have fun, hopefully get a message across, make TV an interesting place, and above all -- BE YOURSELF!.

See you on the tube!

Fade to black *"Bye!"*

The End

REFERENCE SECTION

Public Access Studios

Los Angeles

City of Los Angeles Information and Technology
Agency
120 South San Pedro St., Suite 600
Los Angeles, CA 90012
(213) 485-2866
It used to be Tony, then Terry, now it's Angeles who
is in charge of all Public Access.

All studios have the basics: lights, three cameras,
furniture, plants, full crew, Teleprompter, "Live" Call-
in capability. Most provide two hours to produce --
some only ninety minutes, while others offer up to
three hours.

Total dwelling units range from about 18,000 for
Cox to about 270,000 for Century to over 600,000
for all of MediaOne. This gives you an idea of how
many people could be watching you on just three
stations.

CENTURY COMMUNICATIONS (310) 315-4444
(Channel 77)
2939 Nebraska Ave.
Santa Monica, CA 90404
The airtime schedule (310) 829-1212. "Live" Call-in
number (310) 315-4428 for Saturdays.
Ask for Mosa (intern program) (310) 315-4486, or
Brinda (310) 315-4445 -- both are very good and
creative directors.
If you have a problem, the person who oversees the
department is John Monaghan -- who used to work
in the public access studio -- (310) 315-4400. Just
drop my name!

CENTURY CABLE (323) 255-9881
3037 Roswell St.
Los Angeles, CA 90065
They're out by Dodgers Stadium.
Ask for David Yerena -- again, use my name, he's a
great guy!

MEDIAONE (323) 993-8000
900 N. Cahuenga Blvd.
Hollywood, CA 90038
They used to be known as Continental, and they
have a production van (check out Chapter 16). I
just spoke with Pat who is in programming.

MEDIAONE (310) 822-1575
4223 Glencoe Ave. #C125
Marina Del Rey, CA 90292
One of the newer facilities out towards the beach.

MEDIAONE (310) 216-3525
6314 Arizona Place
Los Angeles, CA 90045

MEDIAONE (818) 353-9304
10625 Plainview Ave. #10
Tujunga, CA 91042-0188

TIME WARNER (818) 998-2266
Programming is (818) 998-2281
9260 Topanga Canyon Blvd.
Chatsworth, CA 91311. (Closed Sun. & Mon.)
Ask for Eric Lee Smith. He worked on some of my
shows as an intern and now runs a studio -- he's
very helpful.

TCI (818) 781-1900
15055 Oxnard St.
Van Nuys, CA 91411
Channel 25, and Leased Access is Channel 26
The Studio Formerly Known as United Artists. I just
spoke to Larry Jones -- he's in charge, very
knowledgeable and a good guy -- again, use my
name. There's also a rumor that Century will be
taking over all of TCI -- cable mergers and stuff.

MEDIAONE (323) 565-2807
1855 W. Manchester Ave.
Los Angeles, CA 90047

MEDIAONE (310) 513-1534 (Channel 41)
605 East "G" Street
Wilmington, CA 90744-6027

BUENAVISION (323) 269-8266 (Channel 6 & 56)
912 North Eastern Ave.
Los Angeles, CA 90063

COX CABLE (310) 377-7207 (Channel 33)
33 Peninsula Center
Rolling Hills Estates, CA 90274

If you live in Los Angeles, the following out of state access studios are helpful for bicycling your shows.

NEW YORK:
Manhattan Neighborhood Network (212) 757-2670
537 W. 59th St.
New York, NY 10019
(Near Columbus Circle)
This is probably the biggest and most popular studio in the country, with four channels!

BronxNet (718) 960-1180
BCAT (Brooklyn) (718) 935-1122
QPTV (Queens) (718) 886-8160
CTV (Staten Island) (718) 727-1414

Problems?
Mayor's Office
City Hall
New York, NY 10007
(212) 788-9600

OUTSIDE NEW YORK CITY:
HAMPTONS (516) 727-6300
LONG ISLAND (718) 358-0900
MAMARONEK (914) 899-9000

OTHER CITIES
Austin Community Access--ACTV (Austin , TX)
(512) 478-8600
1143 Northwestern
Austin, TX 78702
Hot Tip!!! I just spoke to Marion Nickerson in the programming department -- he told me to have you write him a letter and he'll post it for sponsorship -- (a trade off). This means someone, a resident, will air your shows in exchange for airing theirs in your city. Probably one of the coolest places in the

country. I used to air my show there, and they were great to work with. Give them a call.

Access Tucson. (Tucson, AZ) (520) 624-9833
124 E. Broadway
Tucson, ,AZ 85701
Part of Cox Cable (Channel 54 & 56)
Sam Behrend is in charge.
They reach over 90,000 households.

Check out the following two publications in your local library:

BACON'S TV/CABLE DIRECTORY
(800) 972-9252 for research

SRDS TV & CABLE SOURCE QUARTERLY
(847) 375-5000

And check these two cool organizations:

ALLIANCE FOR COMMUNITY VIDEO
(202) 393-2650. Washington, DC
They are a national organization that sponsors a conference, trade show and video festival.

ROCK THE VOTE
www.rockthevote.org

Music Video People

Remember use my name when contacting the music video people. Some people have left their jobs, but the fax should be the same. Fax a letter of inquiry first! Then make a follow up call. The bigger labels distribute the smaller ones. For example: Warner Bros. handles Madonna's Maverick label. The independents work with various labels and their artists. I've included some addresses for the smaller labels and independent video promoters, too. And one last note -- Universal bought Polygram, which owned a bunch of labels. For example, Interscope handles Geffen, A&M and Mercury now.

Note: ALL NUMBERS ARE FAX NUMBERS (unless noted).

A&M Records. Dean, (212) 333-1322 or call (212) 333-1398 Jenay Davis.
Sting, Sheryl Crowe, 10 Speed, God Lives Underwater, Elton John, Ryan Downe.

Aristomedia (Independent Country Music).
They are in Nashville (615) 269-0131
or call toll free 800-947-7071 ext. 142.
P.O. Box 22765
Nashville, TN 37202
They've got some cool country and country rock artists: Derailers, The Mavericks, Dixie Chicks, Shania Twain.

All Access (Independent).
(203) 336-3329

Capitol Records. Bonnie Burkert, (212) 253-3099 or call (212) 253-3000. Www.hollywoodandvine.com
Everclear, Richard Marx, Dandy Warhols, Marcy Playground, Megadeth, Paul McCartney, Beastie Boys, Radiohead, Duran Duran, Sean Lennon, Foo Fighters.

Coventry. (323) 662-0403
phone (323) 662-5031
3272 Descanso Dr.
Los Angeles, CA 90026

Epic Records. (212) 833-8496, fax to Suki.
Oasis, Puff Daddy, Jimmy Ray, Gloria Estefan,
Pearl Jam, Korn. email: Barbara -
rubenstein@sonymusic.com

Geffen Records (310) 274-7331
You might have to go through Interscope.
Beck, Garbage, Crystal Method, The Sundays.

Hollywood (212) 645-9427, fax to Adam.
Queen, Suicide Machines, Fastball, Pretenders.

Interscope (212) 980-7042
Bush, Body Count, Nine Inch Nails, No Doubt,
Marilyn Manson, Smashmouth, Wallflowers, Primus.

Island Records (212) 603-3965
U2, Tom Waits, Grace Jones, All Saints, Harvey
Danger, Pulp, Echo and the Bunnymen,
Cranberries.

Loud Records (310) 358-4581
phone: (310) 358-4550, or ask for Chanel Green
(310) 358-4557
8750 Wilshire Blvd. 2nd Flr.
Beverly Hills, CA 90211
Specialize in Rap artists.

MCA Records (212) 841-8146
Jimi Hendrix, Blink 182, Semisonic, Sublime, The
Nixons, Dance Hall Crashers, Aqua.

Mojo Records Yvonne Garrett (212) 966-9573
phone (212) 966-7334
Reel Big Fish, Cherry Poppin' Daddies, Goldfinger.
1749 14th St., #201
Santa Monica, CA 90404
Local phone:(310) 260-3181

MVP (Independent) (917) 441-7016 Laurie Nocerito
phone:(212) 579-9842

Reprise (818) 840-2424
B- 52, Eric Clapton, Green Day, Depeche Mode.
.
RCA Records (212) 930-4167. Phone: 800-627-
4926, ask for Jeannine Panaccione.
Dave Matthews, Robert Bradley, The Verve Pipe.

Risk Records Steven Cohen (888) 409-2283

Sire Scott Graves phone (212) 253-3931

Sony Music Ent.
Call: (212) 833-8146, ask for Andrea Hall

Telemotion (Independent). Laurel or Aaron.
(323)651-2073 or phone (323) 651-2070
Jewel, Cherry Poppin Daddies, Reel Big Fish.
P.M.B. 286
8424A Santa Monica Blvd
Los Angeles, CA 90069

Trauma Records
15165 Ventura Blvd., Suite 320, Sherman Oaks, CA
91403

V2 Records. (212) 320-8720, or call Stacy Kanter @ (212) 320-8585
14 E. 4th St.
New York, NY 10012

Vapor Records (310) 393-6512 or call Bonnie Levetin @ (310) 393-8442
1460 4th St. #208
Santa Monica, CA 90401
Jonathan Richman

Virgin Records Stephanie Seymour (212) 253-3099 or phone (212) 253-3111. She loves faxes and emails!
David Bowie. Oasis, Janet Jackson, Spice Girls, Smashing Pumpkins, The Verve, Lenny Kravitz, Sneaker Pimps, Rolling Stones, Blur, DC Talk.

Vis-Ability (Independent). Jeff or Julie
(323) 653-0482 or phone 323-658-8766
Marilyn Scott, Everclear, Megadeth, Robbie Robertson, Duran Duran.
email: jamato3836@aol.com
7958 Beverly Blvd.
Los Angeles, CA 90048

Warner Bros. (212) 275-3830
Madonna, Van Halen, Alannis Morrisette, Prodigy, Rod Stewart, Fleetwood Mac, Jane's Addiction, Deftones, Paula Cole, Presidents of the USA.

Wind-up Ent. (212) 251-0779
or phone Kim Burke (212) 251-9665

MUSICAL ORGANIZATIONS

Look on the inside of CD's and cassettes and you'll see either one of the two performing rights organizations:

BMI (Broadcast Music Inc.) PH: (310) 659 9109
FX: (310) 657-6947
8730 Sunset Blvd. 3rd Flr. West
West Hollywood, CA 90069
Song research department for information on copyright holders and publishing companies.

ASCAP (American Society of Composers, Authors and Publishers) PH:(323) 883-1000
FX: (323) 883-1049
7920 Sunset Blvd. #300
Los Angeles, CA 90046
Index department for information on copyright holders and publishing companies.

Songwriters Guild of America, The
PH: (323) 462-1108 FX: (323) 462-5430, or
PH: (201) 867-7603 FX: (201) 867-7535
www.songwriters.org

Note: National Academy of Songwriters has closed and The Songwriters Guild of America is taking in new members. They've been around a long time, but I don't know much about them.

MUSIC

You buy it, you own it, you use it.

Contempo Music PH: (888) 687-4285
P.O. Box 2049
Elmhurst, IL 60126

Davenport Music PH: (800) 951-6666
P.O. Box 25636
Charlotte, NC 28229

Energetic Music PH: (800) 323-2972
P.O. Box 84583
Seattle, WA 98124

Hollywood Edge PH: (800) 292-3755
7080 Hollywood Blvd. Suite 519
Hollywood, CA 90028

QCCS Prods. PH: (800) 345-7286
1350 Chambers St.
Eugene, OR 97402

Royalty Free Music PH: (800) 772-7701

Soper Sound PH: (800) 227-9980
Hear selections over the phone.

Caution: I recently edited a commercial with music
from Killer Tracks in Hollywood, where they had to
do a "needle drop," which means you pay each time

you use it (it airs). It starts to add up and bloat your budget.

Movie Clips:

For getting listed on the International Motion Picture Press Directory. If listed, you can get the Studio Publicity Departments to send you clips automatically.

Motion Pictures Association of America
PH: (818) 995-6600 FX: (818) 382-1799

Requirements:
1) Introductory letter of why you need accreditation plus the length and schedule of your program and size of audience.
2) A letter from the studio where you air your show verifying it's on -- ask the traffic person.
3) VHS cassette of a sample show where you've interviewed some actors and talked about movies.
4) Fill out a questionnaire. You can get it by calling them up and requesting the TV accreditation procedures. PH: (818) 995-6600. Ask for Public Relations. The last person there was Yulia Dashevsky.

Their address is:
15503 Venture Blvd.
Encino, CA 91436

CLUBS, COFFEEHOUSES, etc.

LOS ANGELES AREA

Check with your local Starbucks, Diedrich's, etc, or favorite coffeehouse. Plus, there are a plethora of local papers telling you where to go.

Anastasia'a Asylum PH: (310) 394-7113
1028 Wilshire Blvd.
Santa Monica, CA
When we were dating, my wife and I used to hang out here, sip coffee drinks, and kiss on the sofa.

Common Grounds PH: (818) 882-3666
9250 Reseda Blvd.
Northridge, CA

Highland Grounds PH: (323) 466-1507
742 N. Highland Ave.
Los Angeles, CA
Great place for meeting new people.

Hothouse Cafe PH: (818) 506-7058
12123 Riverside Dr.
North Hollywood, CA

Little Frida's PH: (310) 854-5421
8739 Santa Monica Blvd.
West Hollywood -- this part of L.A. is a concentrated mecca for clubs that are mostly gay and lesbian.

Moondog Cafe PH: (323) 936-4604
7160 Melrose Ave.
West Hollywood, CA

Un-Urban PH: (310) 315-0056
3301 Pico Blvd.
Santa Monica, CA
Right around the corner from the Century Cable
studio.

Borders Books & Music PH: (310) 393-9290
1415 Third St. Promenade
Santa Monica, CA
One of my guests from *Pop Bop!*, Jonatha Brooke,
peformed here in March '99.

World Cafe PH: (310) 392-1661
2820 Main St.
Santa Monica, CA

Troubadour PH: (310) 276-6168
9801 Santa Monica Blvd.
West Hollywood, CA
This is where Elton John made his debut!

Genghis Cohen PH: (323) 653-0640
740 N. Fairfax
West Hollywood, CA
A friend of mine played here many times -- very
good!

Largo PH: (323) 852-1073
432 N. Fairfax
West Hollywood, CA
I worked on a movie with Oscar nominated Actress,
Mare Winningham, who played here -- good place.

Molly Malones Irish Pub PH: (323) 935-1577
575 S. Fairfax
Los Angeles, CA
Cheap cover -- good place to see small bands.

NEW YORK AREA

Check out the *Village Voice*, and of course, your local coffeehouses, no matter what city you live in!

Supplies

I mentioned the following company early on, but you can also ask your studio to recommend a good place that is close to you.

Studio Film and Tape Carole Dean (president)
Michele Mills PH: (323) 466-8101
1215 N. Highland Ave.
Hollywood, CA 90038
800-824-3130 Fax (323) 466-6815
Ask the counter person about any Cable Access Producer discount. Plus, they also have grants available for documentaries -- just ask.

New York
PH: (212) 977-9330 / 800-444-9330
FX: (212) 586-2420
630 Ninth Ave
New York, NY 10036

Chicago
PH: PH: 800 467-0070
FX: (312) 467-0074
110 W. Kinzie St., 1st floor
Chicago, IL 60610

GLOSSARY

Credits: Tells the viewer what your show is called, who the host and guest is and information needed at the end of the show, i.e., sponsors, your phone number, etc.

Demonstration: A type of show where you teach people how to do something -- cooking, haircutting, dance, karate, exercise, etc. (Chapter 3).

Detail Switch: Located on most cameras to soften the edges of the video image -- you/guest.

Diffused Lighting: A gel or piece of special paper they put over the white lights to soften the intensity.

Editing: Putting segments of video footage -- including audio -- together in some kind of order (Chapter 17).

Interning: Working at a studio for free while learning all about production (Chapter 18).

Just You: A type of show where you do it alone (no guests) (Chapter 2).

Leased Access: Basically, Public Access with commercials.

Mastercontrol: No, not Houston -- it is where they air tapes from.

Performance: A type of show where you entertain -- musical, acting, puppets (Chapter 4).

Public Access:. Located at the local cable company. Where you go to produce and star in your show, then air it. (a/k/a Cable Access)

Rundown Sheet: This piece of paper tells the director and crew exactly what you are doing for your show -- including, how long your opening is, to how many guests you're interviewing, to how many roll-ins you have.

Roll-ins: Video clip(s) footage you want to show during your show. With or without sound -- sometimes you'll want to talk about the footage as you watch it.

Studio: Where you actually tape your show. It should have three cameras, a staging area, lights, microphones, a control room filled with all kinds of machines.

Talk Show: Two chairs, a table, some plants, with a host and a guest sitting and talking (Chapter 2).

TBC (Time Based Corrector): A special effects unit -- Strobe, Freeze, Mosaic and Sepia. Also used for adjusting the video level (bright), set up (the black level), Chroma and Hue.

3/4" Videotape: The general format required to record and broadcast your show.

Variety: Some combination of the four previous show types: Talk, Demonstration, Performance and Just you (Chapter 5).

INDEX

A
AIDS Vision, 11
Alcohol, 91
America's Most Wanted, 113
APLA, 208
Art Fein's Poker Party, 15
ASCAP, 98
Ask Livia Live, 206
Ask Mr. Traffic, 52, 204

B
Barbra, 38
Basic Judaism, 52
Billboard (Magazine), 116, 196
Bicycling, 165
Block, Dr. Susan, 83, 167
BMI, 98
Bogdanovich, Peter, 114
Brown, Mason, 41
Budget, 134

C
Cancellations, 87
Carl Bradley Pitches TV Show Ideas, 14
Celestine Prophecies, 112